William P. Spence

# ENGINEERING GRAPHICS

## WORKBOOK
### Volume II
### Second Edition

Prentice Hall, Inc., Englewood Cliffs, New Jersey 07632

Editorial/production supervision and
    interior design: *Maureen Lopez*
Manufacturing buyer: *Gordon Osbourne*

Printed in the United States of America

10  9  8  7  6  5  4  3  2  1

ISBN 0-13-277955-2   025

Prentice-Hall International (UK) Limited, *London*
Prentice-Hall of Australia Pty. Limited, *Sydney*
Prentice-Hall Canada Inc., *Toronto*
Prentice-Hall Hispanoamericana, S.A., *Mexico*
Prentice-Hall of India Private Limited, *New Delhi*
Prentice-Hall of Japan, Inc., *Tokyo*
Simon & Schuster Asia Pte. Ltd., *Singapore*
Editora Prentice-Hall do Brasil, Ltda., *Rio de Janeiro*

# CONTENTS

# INTRODUCTION

The problems in this workbook are closely related to the text, Engineering Graphics by William P. Spence, published by Prentice-Hall, Inc. However, since they cover the basic principles of engineering graphics, they can be used with any similar graphics text.

The problems are arranged following the outline of the text. However, the instructor can easily alter the order in which they are used so as to meet his/her immediate needs.

Some of the problems are in metric units while others are in customary units. During this period when industry is gradually moving into metrics engineers and technologists must be able to use either with equal ease. This workbook provides the instructor the opportunity to give students an understanding of and experiences with both systems.

Many of the problems are partially laid out on the sheets. This will reduce the time spent on routine layouts of the problems. By devoting as much of the limited time that is available to problem solving a greater number of areas can be studied and learning opportunities broadened.

It is important for engineers and technologists to be able to letter in a clear, professional manner. Part of the evaluation of your drawing by your co-workers and supervisors is the impression given by the lettering. While a poorly lettered drawing may be drawn accurately, the poor lettering could introduce the possibility of error when the drawing is used. The first sheets in this workbook provide for lettering practice. It is suggested that they be used at various times throughout the course rather than completing all of them first. It is a good idea to give the students time to practice lettering by requiring that class notes and reports be hand lettered. Later lettering practice sheets should show improvement.

Sketching is another skill needed by the engineer and technologist. After studying sketching techniques , practice is necessary to develop this skill. It is a plus if a person can quickly produce a clean, well made sketch of an idea or proposed solution. This is important to recording ideas and is an impressive engineering skill. Some of the sheets in this workbook have printed grids to help with the sketching experience. It is suggested that students sketch solutions to the problems before they begin to draw them on the sheets provided. This not only develops sketching skills but teaches an orderly way to approach a problem solving situation.

As students approach problems they must learn to read the directions carefully. After producing a sketch of the solution they need to recheck it with the directions. After completing the finished drawing they need to reexamine the entire thing to see that the specified conditions have been satisfied.

This workbook will provide a thorough experience in the basic areas of engineering graphics. Those who pay close attention and work diligently to find the solutions will have mastered the basics needed for engineers and technologists.

DRAW EACH LETTER IN THE SPACE PROVIDED. USE AN H OR F PENCIL. OBSERVE THE FORM
AND PROPORTION OF EACH LETTER.

A       K

B       L

C       M

D       N

E       O

F       P

G       Q

H       R

I       S

J       T

U      V      W

X      Y      Z

LETTER EACH NUMERAL IN THE SPACE PROVIDED. USE AN H OR F PENCIL. OBSERVE THE FORM
AND PROPORTIONS OF EACH NUMERAL.

1                6

2                7

3                8

4                9

5                0

$\frac{1}{4}$             $\frac{3}{8}$            $\frac{5}{16}$

.50           6.2

LETTER THE DIMENSIONS AND NOTES
BELOW ON THE DRAWING AT THE
RIGHT. USE $\frac{1}{8}$" VERTICAL CAPITAL
GOTHIC LETTERS.

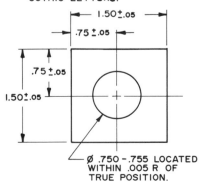

1.50±.05
.75±.05
.75±.05
1.50±.05
Ø .750 –.755 LOCATED
WITHIN .005 R OF
TRUE POSITION.

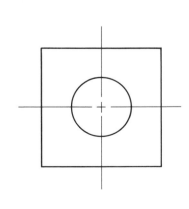

| ENGINEERING GRAPHICS | NAME ⹀ CLASS ⹀ | SECTION ⹀ | DATE ⹀ | GRADE | DRAWING NO. 2 |

ENGINEERS INVOLVED IN RESEARCH ARE  ATTEMPTING TO APPLY KNOWN PRINCIPLES

TO SOLVE EXISTING PROBLEMS OR TO DISCOVER NEW KNOWLEDGE.   THIS

IS USUALLY A PRACTICAL TYPE OF RESEARCH THAT CAN INVOLVE MANY

THINGS,   SUCH AS THE USE OF MATERIALS OR THE DEVELOPMENT OF A

NEW PROCESS.   THIS TYPE OF WORK IS OFTEN SLOW TO SHOW RESULTS.   WHILE

SCIENTIFIC KNOWLEDGE IS THE BASIS FOR RESEARCH,  MANY ATTEMPTS AT

SOLUTIONS ARE BY TRIAL AND ERROR.  RESEARCH ENGINEERS ADD TO THE

KNOWLEDGE THAT CAN BE USED BY OTHERS.   THEY USUALLY WORK ON

RESEARCH TEAMS WHICH HAVE PEOPLE HAVING A VARIETY OF BACKGROUNDS.

**LETTERING PRACTICE**

| ENGINEERING | NAME 		| | | GRADE | DRAWING NO. |
| GRAPHICE | CLASS 	SECTION 	DATE | | | | **3** |

ENGINEERING DESIGN IS THE PROCESS OF ARRIVING AT A PRACTICAL AND ECONOMICAL

SOLUTION TO A PROBLEM.  IT MAY INVOLVE DEVELOPING A NEW PRODUCT OR

IMPROVING AN EXISTING PRODUCT.  IT MAY BE THE DEVELOPMENT OF A NEW

SYSTEM, SUCH AS A MEANS TO REMOVE POLLUTION IN THE AIR IN A

MANUFACTURING SITUATION.  THIS PROBLEM-SOLVING PROCESS CAN INCLUDE

A WIDE VARIETY OF ACTIVITIES, INCLUDING THE USE OF COMPUTERS,

MATHEMATICS, DRAFTING CONSULTANTS, TESTING AND ANY OTHER ACTIVITY

NEEDED TO BRING THE PROJECT TO A SATISFACTORY CONCLUSION.  THE DESIGN

PROCESS INVOLVES MANY PEOPLE WORKING TOGETHER TOWARD A SOLUTION.

**LETTERING PRACTICE**

| ENGINEERING GRAPHICS | NAME ▭ | | | GRADE | DRAWING NO. |
|---|---|---|---|---|---|
| | CLASS ▭ | SECTION ▭ | DATE ▭ | | 4 |

SURFACES AT RIGHT ANGLES TO A DATUM AXIS

W 8 X 35 STEEL BEAM

.250 TAPER ON DIAMETER PER INCH OF LENGTH

$3\frac{1}{2}$" LALLY COLUMN

HARDEN ALL OVER TO ROCKWELL C 40-46.
REMOVE BURRS AND SHARP EDGES TO .030 R.

.250-.245 DIA. HOLE 1.00 DEEP

.125-27 DRYSEAL NPTF

NOTE: ALL FILLETS AND ROUNDS .35 R.

M 10 X 1.5 - 50 DEEP

96 DP STRAIGHT KNURL 1.45 MIN. DIA AFTER KNURLING

$\frac{1}{2}$ X $2\frac{1}{2}$ HEX HD BOLT

NO. 4 AMERICAN NATIONAL STD. TAPER

SHEATHING $\frac{3}{8}$ PLYWOOD

M 8 X 1.25 X 20 RD HD CAP SCREW

CSK 90° X .82 DIA.

$\frac{1}{4}$ - 28 UNF - 2A X $\frac{1}{2}$ FLAT HD MACH SCR.

CAST FINISH FILLETS AND ROUNDS

SURFACE ROUGHNESS ⌄125 UNLESS OTHERWISE NOTED.

∅ .62 1.00 C'BORE .75 DEEP

LETTERING PRACTICE

| ENGINEERING GRAPHICS | NAME | | | | GRADE | DRAWING NO. |
|---|---|---|---|---|---|---|
| | CLASS | SECTION | | DATE | | 5 |

COMPLETE THE TITLE BLOCK
USING THE INFORMATION BESIDE
EACH LINE.

SCHMILLE TOOL CO.

ROCHESTER, NEW YORK

TOOL HOLDER BRACKET

DR. BY: R. SENTE   TR. BY: J. JONES

CR. BY: W. TREE   APP. BY: B. POLE

SCALE: 1:2   5-16-19   SHEET 2

BELOW ARE DATA SHOWING THE ANNUAL
PRODUCTION OF STEEL PRODUCTS.
REARRANGE THESE FROM HIGH TO LOW
AND LETTER THEM IN THE SPACE TO THE
RIGHT.  USE $\frac{3}{16}$" LETTERS SPACED $\frac{1}{8}$" APART.

PRODUCTION FIGURES

| PRODUCT | NET TONS (THOUSANDS) |
|---|---|
| SHEETS AND STRIP | 35,260 |
| BARS | 12,165 |
| PLATES | 8,180 |
| TUBULAR PRODUCTS | 7,225 |
| TIN MILL PRODUCTS | 6,000 |
| STRUCTURAL SHAPES | 5,580 |
| WIRE AND WIRE PRODUCTS | 4,625 |
| SEMIFINISHED SHAPES | 1,275 |

DIMENSION THE FIGURES BELOW USING $\frac{1}{8}$" (3 mm) LETTERS.  ALWAYS USE GUIDELINES.

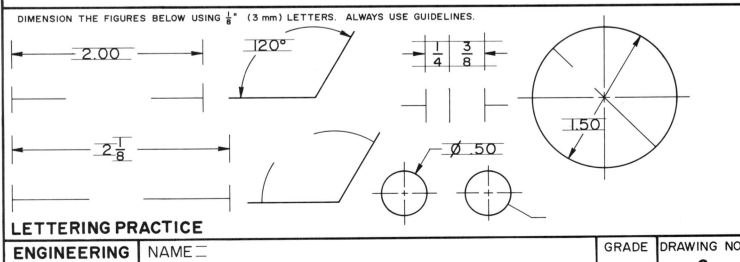

**LETTERING PRACTICE**

| **ENGINEERING** | NAME ⁼ | | | GRADE | DRAWING NO. |
| **GRAPHICS** | CLASS ⁼ | SECTION ⁼ | DATE ⁼ | | **6** |

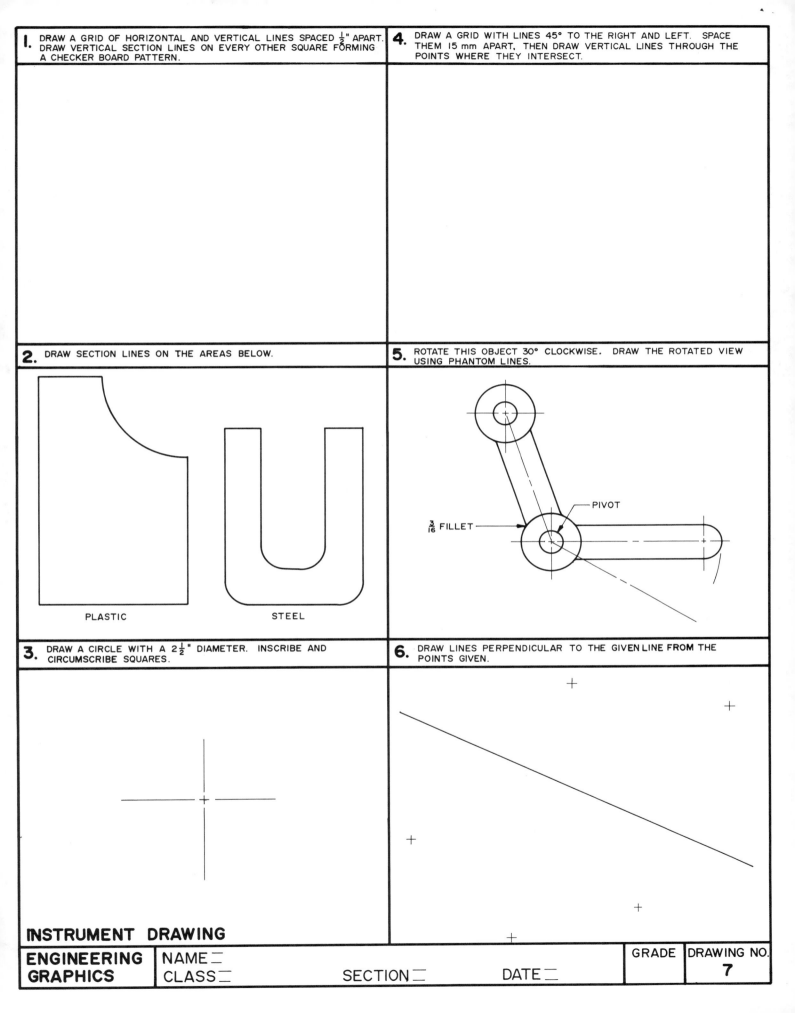

**1.** DRAW A GRID OF HORIZONTAL AND VERTICAL LINES SPACED $\frac{1}{2}$" APART. DRAW VERTICAL SECTION LINES ON EVERY OTHER SQUARE FORMING A CHECKER BOARD PATTERN.

**4.** DRAW A GRID WITH LINES 45° TO THE RIGHT AND LEFT. SPACE THEM 15 mm APART, THEN DRAW VERTICAL LINES THROUGH THE POINTS WHERE THEY INTERSECT.

**2.** DRAW SECTION LINES ON THE AREAS BELOW.

PLASTIC

STEEL

**5.** ROTATE THIS OBJECT 30° CLOCKWISE. DRAW THE ROTATED VIEW USING PHANTOM LINES.

PIVOT

$\frac{3}{16}$ FILLET

**3.** DRAW A CIRCLE WITH A $2\frac{1}{2}$" DIAMETER. INSCRIBE AND CIRCUMSCRIBE SQUARES.

**6.** DRAW LINES PERPENDICULAR TO THE GIVEN LINE FROM THE POINTS GIVEN.

**INSTRUMENT DRAWING**

**ENGINEERING GRAPHICS**

NAME

CLASS

SECTION

DATE

GRADE

DRAWING NO.

**7**

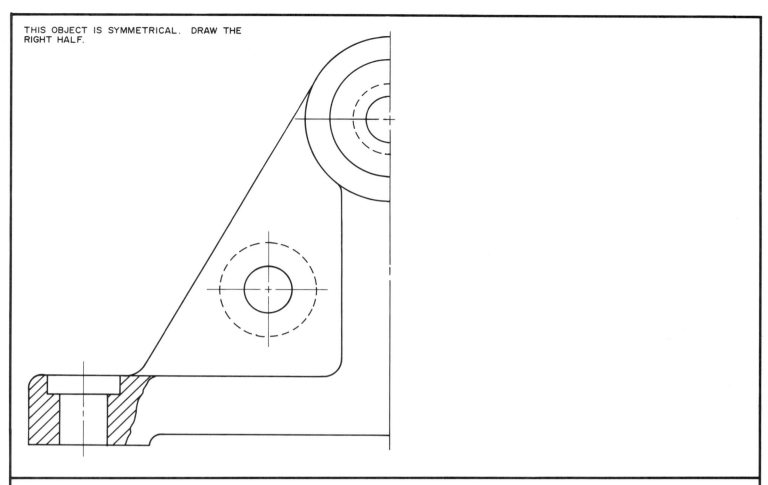

DRAW THIS OBJECT IN THE SPACE AT THE RIGHT
TWICE AS LARGE AS SHOWN. MARK EACH
TANGENT POINT.

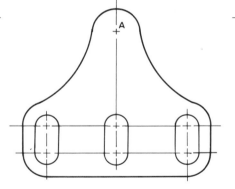

# INSTRUMENT DRAWING

| ENGINEERING<br>GRAPHICS | NAME⁼<br>CLASS⁼ | SECTION⁼ | DATE⁼ | GRADE | DRAWING NO.<br>**8** |
|---|---|---|---|---|---|

IN THE SPACE BELOW DRAW AND DIMENSION THIS STEEL
BEAM. DRAW IT TWICE AS LARGE AS SHOWN.

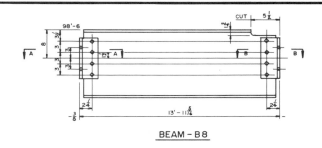

BEAM – B 8

IN THE SPACE BELOW DRAW AND DIMENSION THE FLAT ROOF
CORNICE DETAIL. DRAW IT THREE TIMES AS LARGE AS
SHOWN.

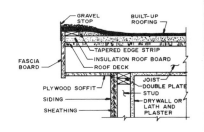

# INSTRUMENT DRAWING

| ENGINEERING GRAPHICS | NAME ⹀ CLASS ⹀ | SECTION ⹀ | DATE ⹀ | GRADE | DRAWING NO 9 |

# READING AN ARCHITECT'S SCALE

MEASURE THE DISTANCES INDICATED ON THE DRAWING AT
THE RIGHT FOR EACH OF THE SCALES LISTED BELOW.
RECORD THE DISTANCES IN THE SPACES PROVIDED.

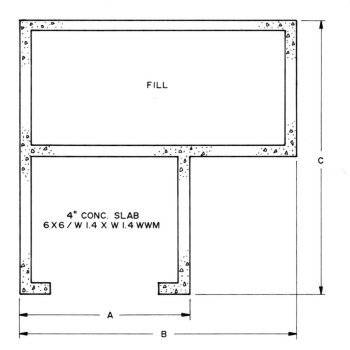

| SCALE | A | B | C |
|-------|---|---|---|
| $\frac{1}{4}$" = 1'-0" | | | |
| $\frac{3}{8}$" = 1'-0" | | | |
| $\frac{1}{2}$" = 1'-0" | | | |
| $\frac{3}{4}$" = 1'-0" | | | |
| $1\frac{1}{2}$" = 1'-0" | | | |

FILL

4" CONC. SLAB
6 X 6 / W 1.4 X W 1.4 WWM

# READING AN ENGINEER'S CHAIN SCALE

MEASURE THE DISTANCES INDICATED IN THE DRAWING
BELOW FOR EACH SCALE LISTED. RECORD THE DISTANCES
IN THE SPACES PROVIDED.

| SCALE | A | B | C | D |
|-------|---|---|---|---|
| 1" = 10.0' | | | | |
| 1" = 20.0' | | | | |
| 1" = 40.0' | | | | |
| 1" = 60.0' | | | | |
| 1" = 300.0' | | | | |

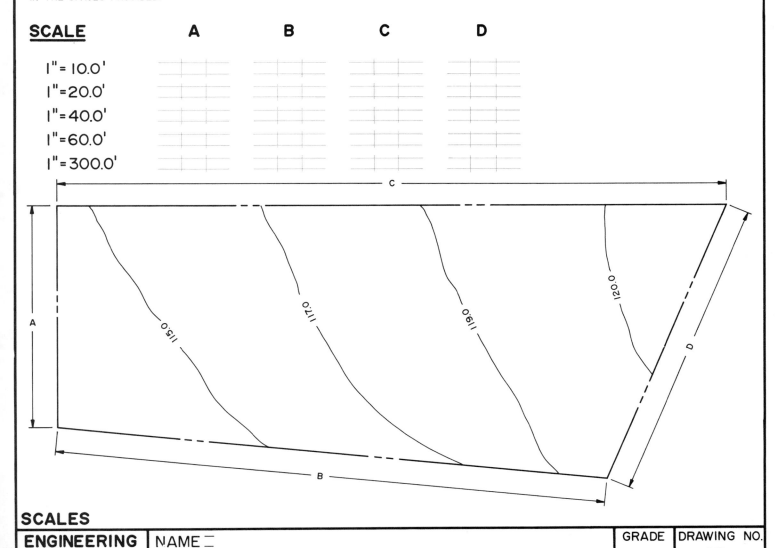

115.0    117.0    119.0    120.0

## SCALES

| ENGINEERING GRAPHICS | NAME = <br> CLASS = | SECTION = | DATE = | GRADE | DRAWING NO. 10 |
|---|---|---|---|---|---|

# READING A METRIC SCALE

MEASURE THE DISTANCES INDICATED ON THE DRAWING AT THE RIGHT FOR EACH OF THE SCALES LISTED BELOW. RECORD THE DISTANCES IN THE SPACES PROVIDED.

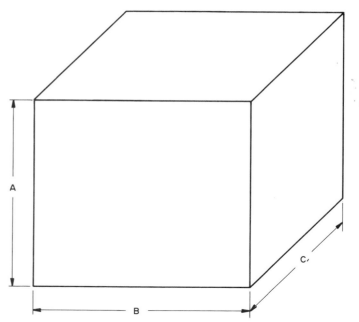

## LINEAR MEASUREMENT

| METRIC SCALE | A | B | C |
|---|---|---|---|
| 1:1 | | | |
| 1:2 | | | |
| 1:5 | | | |
| 1:10 | | | |
| 2:1 | | | |
| 5:1 | | | |

## AREA

| SCALES | A | X | B | = | mm² |
|---|---|---|---|---|---|
| 1:2 | | X | | = | |
| 1:5 | | X | | = | |
| 1:10 | | X | | = | |
| 5:1 | | X | | = | |

## VOLUME

| SCALE | A | X | B | X | C | = | mm³ |
|---|---|---|---|---|---|---|---|
| 1:2 | | X | | X | | = | |
| 1:5 | | X | | X | | = | |
| 1:10 | | X | | X | | = | |
| 2:1 | | X | | X | | = | |

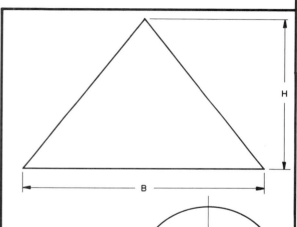

## AREA OF A TRIANGLE

| SCALE | $\frac{B}{2}$ | X | H | = | mm² |
|---|---|---|---|---|---|
| 1:1 | | X | | = | |
| 1:2 | | X | | = | |
| 1:5 | | X | | = | |
| 2:1 | | X | | = | |

## CIRCUMFERENCE OF A CIRCLE

| | $\pi$ | X | D | = | mm |
|---|---|---|---|---|---|
| | | X | | = | |
| | | X | | = | |
| | | X | | = | |
| | | X | | = | |

1:1

10    20    30    40    50    60    70    80    90    100

**SCALES**    **METRIC SCALE – MILLIMETERS**

| ENGINEERING GRAPHICS | NAME⁼ CLASS⁼ | SECTION⁼ | DATE⁼ | GRADE | DRAWING NO. |
|---|---|---|---|---|---|
| | | | | | 11 |

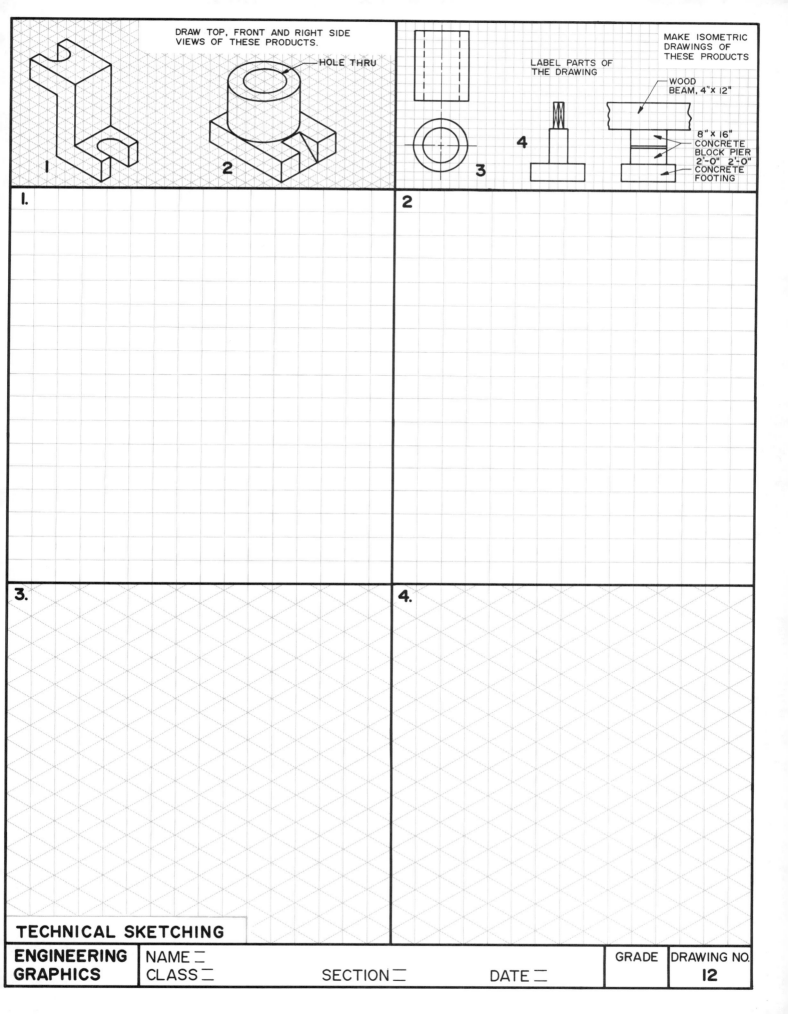

DRAW TOP, FRONT AND RIGHT SIDE VIEWS OF THESE PRODUCTS.

HOLE THRU

1

2

LABEL PARTS OF THE DRAWING

MAKE ISOMETRIC DRAWINGS OF THESE PRODUCTS

WOOD BEAM, 4"x 12"

8"x 16" CONCRETE BLOCK PIER 2'-0" 2'-0"
CONCRETE FOOTING

3

4

1.

2

3.

4.

**TECHNICAL SKETCHING**

**ENGINEERING GRAPHICS**

NAME ⁼
CLASS ⁼          SECTION ⁼          DATE ⁼

GRADE

DRAWING NO.
12

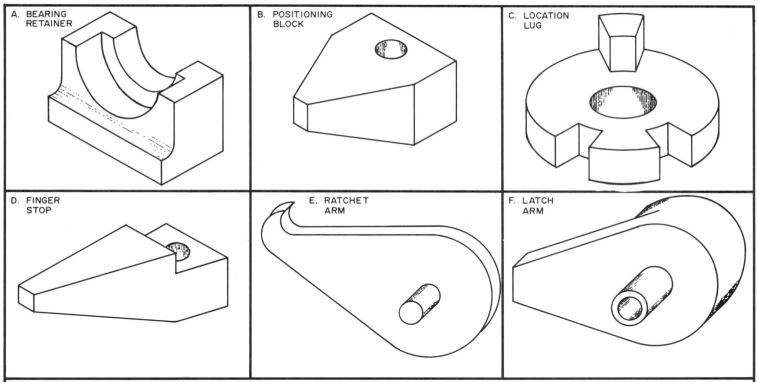

A. BEARING RETAINER

B. POSITIONING BLOCK

C. LOCATION LUG

D. FINGER STOP

E. RATCHET ARM

F. LATCH ARM

SKETCH THE ABOVE PRODUCTS AS PICTORIAL OR MULTIVIEW DRAWINGS AS ASSIGNED BY THE INSTRUCTOR. ESTIMATE THE SIZES. KEEP THE VIEWS IN PROPORTION. USE THE BACK SIDES OF EARLIER SHEETS AS ADDITIONAL DRAWING PAGES.

## TECHNICAL SKETCHING

| ENGINEERING GRAPHICS | NAME ⹀ CLASS⹀ | SECTION⹀ | DATE⹀ | GRADE | DRAWING NO. 13 |

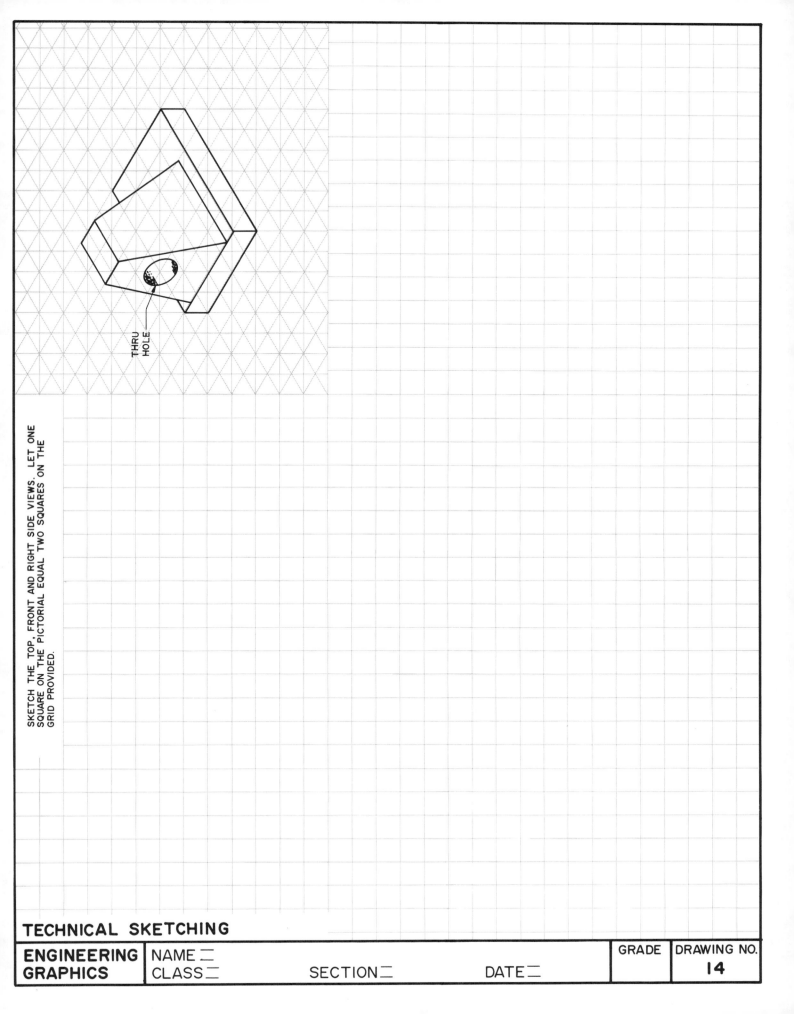

THRU
HOLE

SKETCH THE TOP, FRONT AND RIGHT SIDE VIEWS. LET ONE SQUARE ON THE PICTORIAL EQUAL TWO SQUARES ON THE GRID PROVIDED.

**TECHNICAL SKETCHING**

| ENGINEERING GRAPHICS | NAME ⎓ | | | GRADE | DRAWING NO. |
|---|---|---|---|---|---|
| | CLASS ⎓ | SECTION ⎓ | DATE ⎓ | | **14** |

**1.** DRAW A REGULAR POLYGON WITH SEVEN SIDES. EACH SIDE IS EQUAL TO A-B.

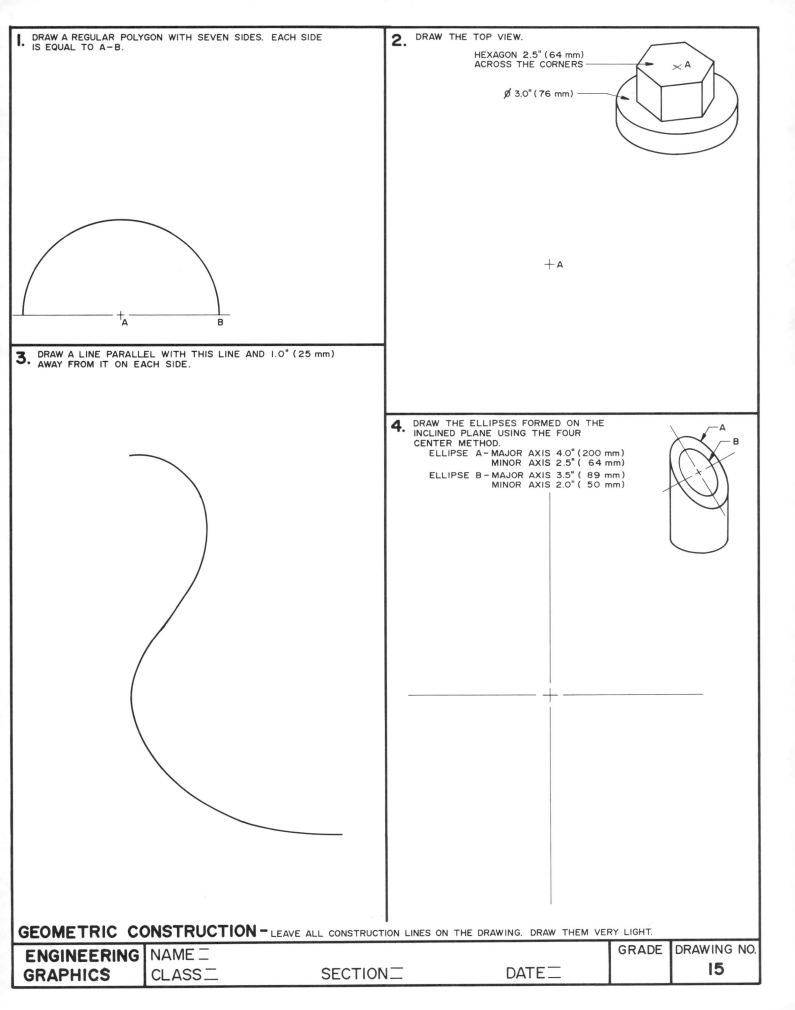

+A          B

**2.** DRAW THE TOP VIEW.

HEXAGON 2.5" (64 mm) ACROSS THE CORNERS

Ø 3.0" (76 mm)

×A

+A

**3.** DRAW A LINE PARALLEL WITH THIS LINE AND 1.0" (25 mm) AWAY FROM IT ON EACH SIDE.

**4.** DRAW THE ELLIPSES FORMED ON THE INCLINED PLANE USING THE FOUR CENTER METHOD.
  ELLIPSE A – MAJOR AXIS 4.0" (200 mm)
              MINOR AXIS 2.5" (64 mm)
  ELLIPSE B – MAJOR AXIS 3.5" (89 mm)
              MINOR AXIS 2.0" (50 mm)

A
B

**GEOMETRIC CONSTRUCTION –** LEAVE ALL CONSTRUCTION LINES ON THE DRAWING. DRAW THEM VERY LIGHT.

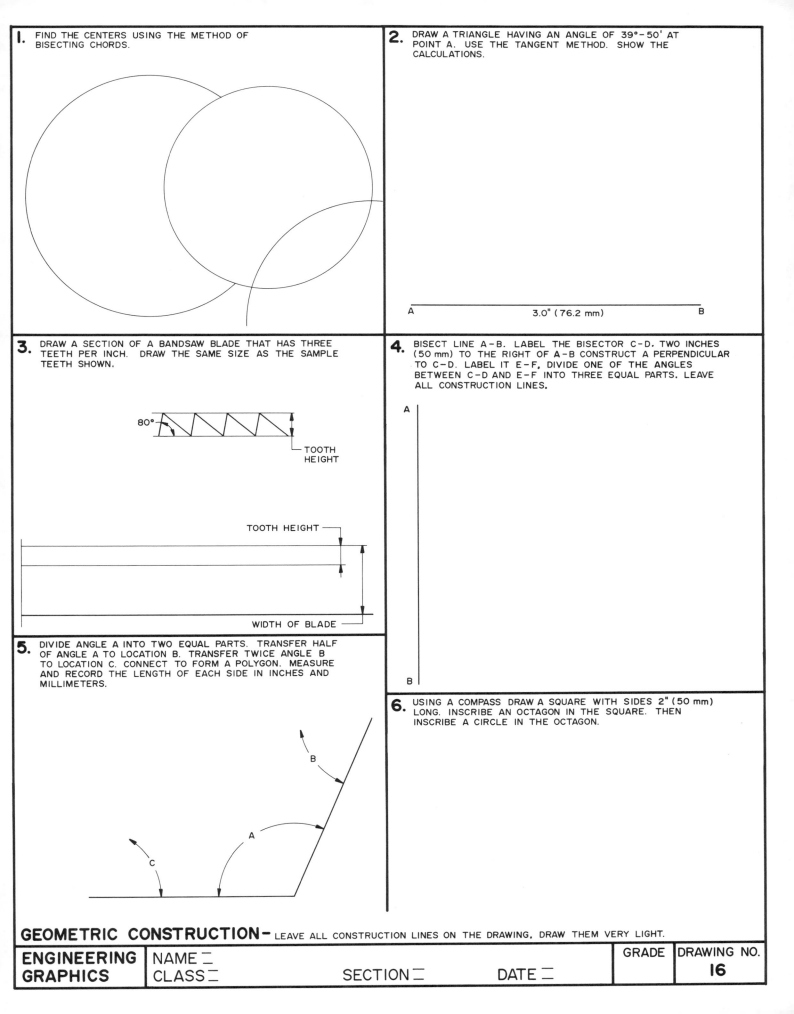

1. FIND THE CENTERS USING THE METHOD OF BISECTING CHORDS.

2. DRAW A TRIANGLE HAVING AN ANGLE OF 39°-50' AT POINT A. USE THE TANGENT METHOD. SHOW THE CALCULATIONS.

A         3.0" (76.2 mm)         B

3. DRAW A SECTION OF A BANDSAW BLADE THAT HAS THREE TEETH PER INCH. DRAW THE SAME SIZE AS THE SAMPLE TEETH SHOWN.

80°

TOOTH HEIGHT

TOOTH HEIGHT

WIDTH OF BLADE

4. BISECT LINE A-B. LABEL THE BISECTOR C-D. TWO INCHES (50 mm) TO THE RIGHT OF A-B CONSTRUCT A PERPENDICULAR TO C-D. LABEL IT E-F, DIVIDE ONE OF THE ANGLES BETWEEN C-D AND E-F INTO THREE EQUAL PARTS. LEAVE ALL CONSTRUCTION LINES.

A

B

5. DIVIDE ANGLE A INTO TWO EQUAL PARTS. TRANSFER HALF OF ANGLE A TO LOCATION B. TRANSFER TWICE ANGLE B TO LOCATION C. CONNECT TO FORM A POLYGON. MEASURE AND RECORD THE LENGTH OF EACH SIDE IN INCHES AND MILLIMETERS.

B

A

C

6. USING A COMPASS DRAW A SQUARE WITH SIDES 2" (50 mm) LONG. INSCRIBE AN OCTAGON IN THE SQUARE. THEN INSCRIBE A CIRCLE IN THE OCTAGON.

**GEOMETRIC CONSTRUCTION-** LEAVE ALL CONSTRUCTION LINES ON THE DRAWING. DRAW THEM VERY LIGHT.

| **ENGINEERING GRAPHICS** | NAME ⁼ CLASS ⁼ | SECTION ⁼ | DATE ⁼ | GRADE | DRAWING NO. 16 |

**1.** DRAW A PARABOLA HAVING A RISE OF 3" (150 mm) AND A SPAN OF 4" (200 mm).

SPAN

RISE

**2.** DRAW A HYPERBOLA USING THE VERTICES, $V_1$ AND $V_2$, AND FOCI, $F_1$ AND $F_2$, GIVEN.

$F_1$

$V_1$

$V_2$

$F_2$

**3.** DRAW A RIGHT-HAND CYLINDRICAL HELIX HAVING A LEAD OF 2" (50 mm).

**4.** DRAW AN INVOLUTE OF THE SQUARE GIVEN BELOW. CONTINUE DRAWING UNTIL THE SPACE IS FILLED.

**GEOMETRIC CONSTRUCTION** LEAVE ALL CONSTRUCTION LINES ON THE DRAWING. DRAW THEM VERY LIGHT.

| ENGINEERING GRAPHICS | NAME ⁓ CLASS ⁓ | SECTION ⁓ | DATE ⁓ | GRADE | DRAWING NO. 17 |

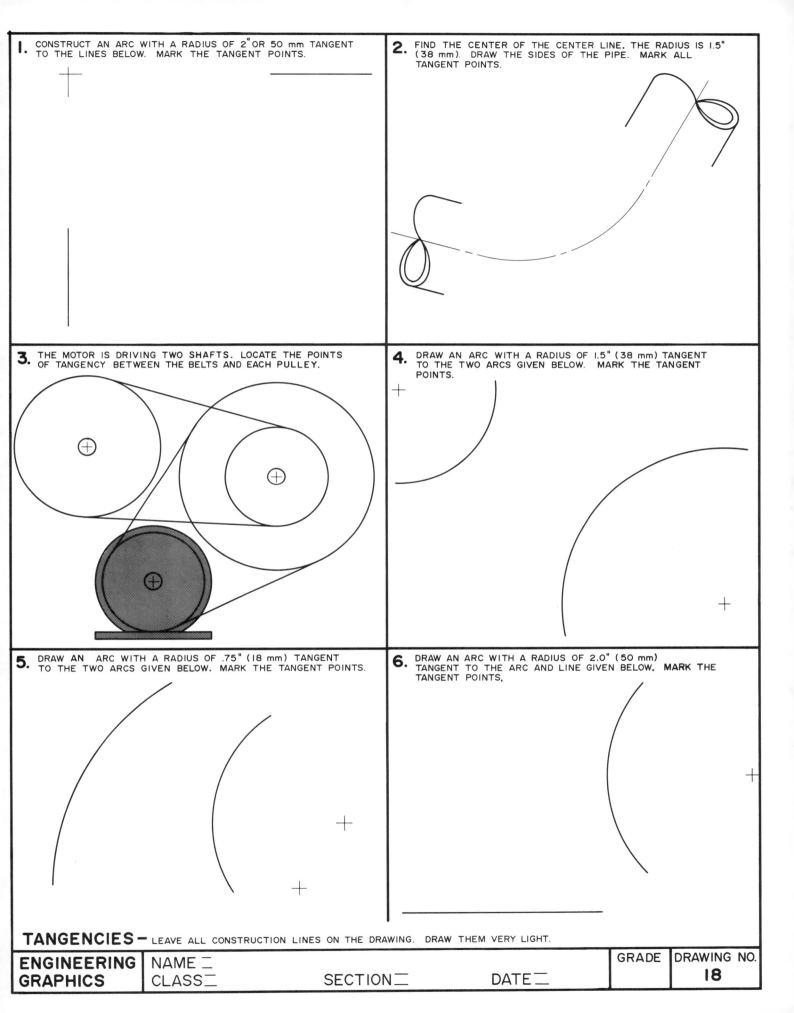

**1.** CONSTRUCT AN ARC WITH A RADIUS OF 2" OR 50 mm TANGENT TO THE LINES BELOW. MARK THE TANGENT POINTS.

**2.** FIND THE CENTER OF THE CENTER LINE. THE RADIUS IS 1.5" (38 mm). DRAW THE SIDES OF THE PIPE. MARK ALL TANGENT POINTS.

**3.** THE MOTOR IS DRIVING TWO SHAFTS. LOCATE THE POINTS OF TANGENCY BETWEEN THE BELTS AND EACH PULLEY.

**4.** DRAW AN ARC WITH A RADIUS OF 1.5" (38 mm) TANGENT TO THE TWO ARCS GIVEN BELOW. MARK THE TANGENT POINTS.

**5.** DRAW AN ARC WITH A RADIUS OF .75" (18 mm) TANGENT TO THE TWO ARCS GIVEN BELOW. MARK THE TANGENT POINTS.

**6.** DRAW AN ARC WITH A RADIUS OF 2.0" (50 mm) TANGENT TO THE ARC AND LINE GIVEN BELOW. MARK THE TANGENT POINTS.

**TANGENCIES –** LEAVE ALL CONSTRUCTION LINES ON THE DRAWING. DRAW THEM VERY LIGHT.

| ENGINEERING GRAPHICS | NAME ‒ CLASS ‒ | SECTION ‒ | DATE ‒ | GRADE | DRAWING NO. 18 |

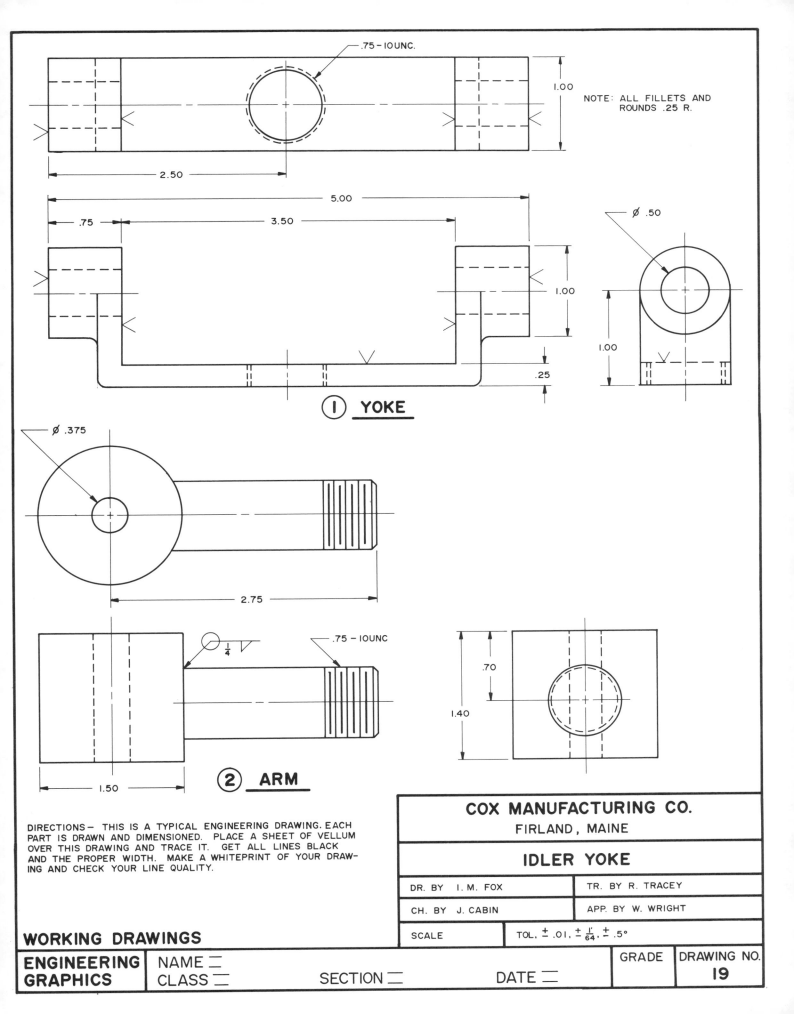

.75-10UNC.

1.00

NOTE: ALL FILLETS AND
ROUNDS .25 R.

2.50

5.00

.75

3.50

1.00

Ø .50

1.00

.25

① **YOKE**

Ø .375

2.75

□ ¼ ▽

.75-10UNC

.70

1.40

1.50

② **ARM**

| COX MANUFACTURING CO. | |
|---|---|
| FIRLAND, MAINE | |
| **IDLER YOKE** | |
| DR. BY I. M. FOX | TR. BY R. TRACEY |
| CH. BY J. CABIN | APP. BY W. WRIGHT |
| SCALE | TOL. ± .01, ± $\frac{1}{64}$, ± .5° |

**WORKING DRAWINGS**

| **ENGINEERING** **GRAPHICS** | NAME ⎯ CLASS ⎯ | SECTION ⎯ | DATE ⎯ | GRADE | DRAWING NO. **19** |
|---|---|---|---|---|---|

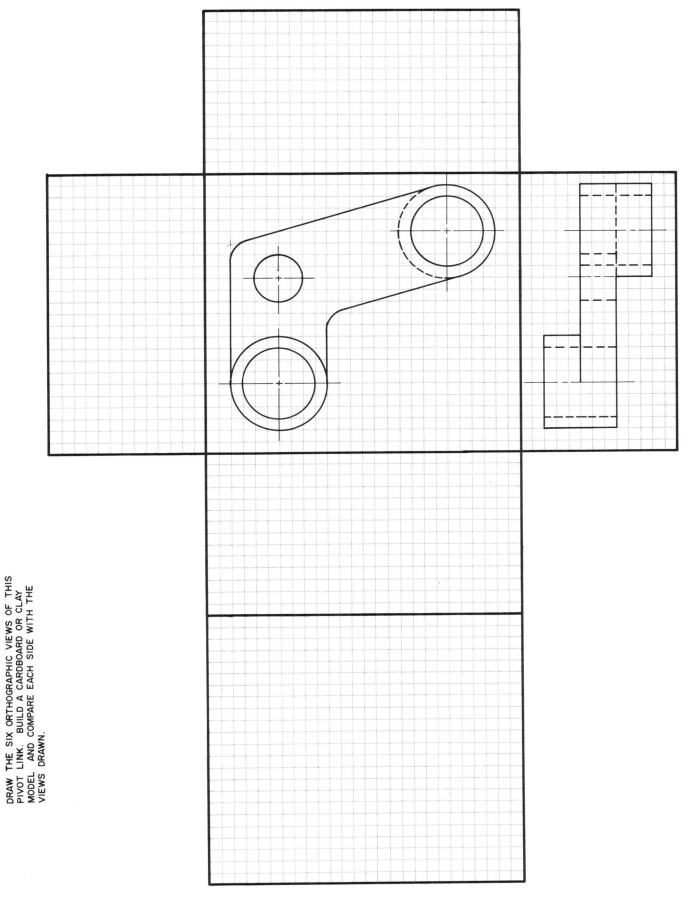

## ORTHOGRAPHIC PROJECTION

| ENGINEERING | NAME ⎓ | | | GRADE | DRAWING NO. |
|---|---|---|---|---|---|
| GRAPHICS | CLASS ⎓ | SECTION ⎓ | DATE ⎓ | | 20 |

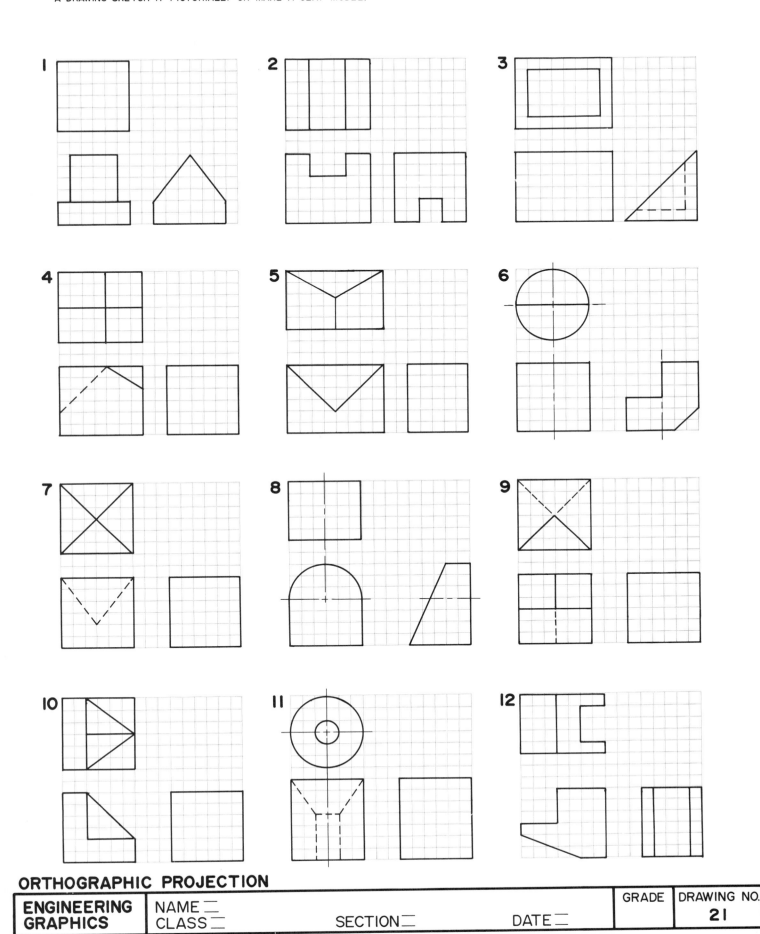

# ORTHOGRAPHIC PROJECTION

| ENGINEERING GRAPHICS | NAME ⬓ CLASS ⬓ | SECTION ⬓ | DATE ⬓ | GRADE | DRAWING NO. 21 |
|---|---|---|---|---|---|

# ORTHOGRAPHIC PROJECTION

| ENGINEERING GRAPHICS | NAME ⹀ CLASS ⹀ | SECTION ⹀ | DATE ⹀ | GRADE | DRAWING NO. 22 |
|---|---|---|---|---|---|

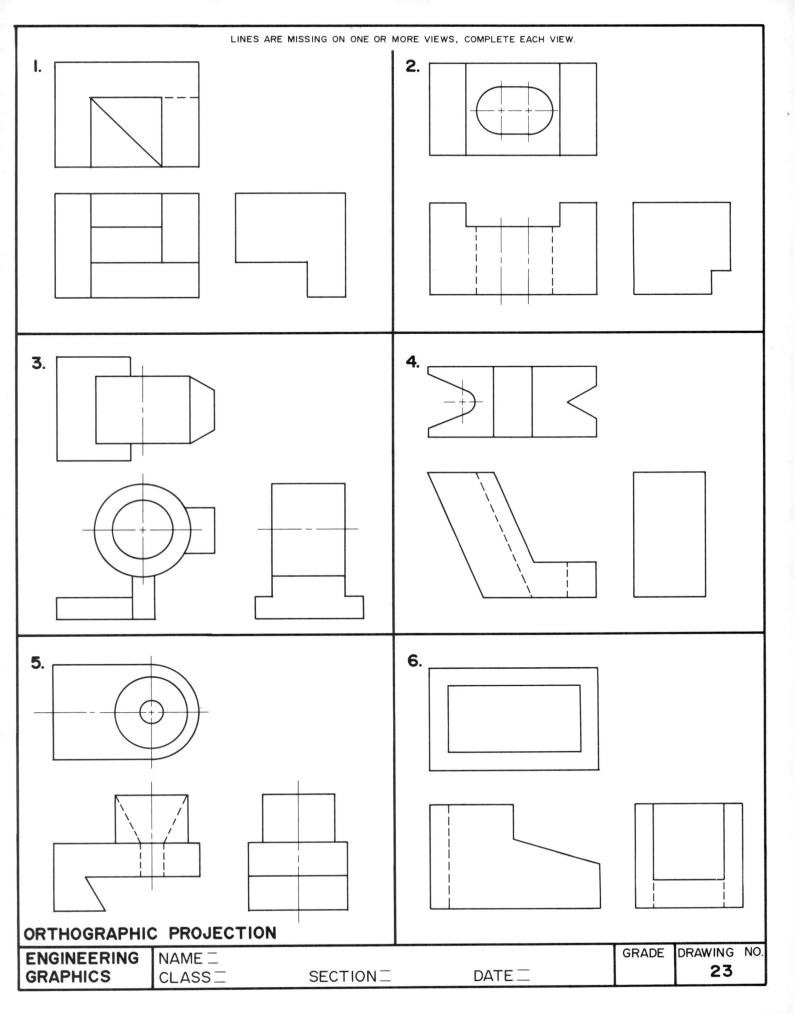

**ORTHOGRAPHIC PROJECTION**

| ENGINEERING GRAPHICS | NAME ⹀ | | GRADE | DRAWING NO. |
|---|---|---|---|---|
| | CLASS ⹀ SECTION ⹀ DATE ⹀ | | | **23** |

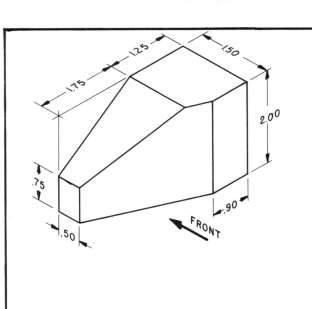

1. DRAW THE TOP, FRONT AND LEFT SIDE VIEWS.

1.75  1.25  1.50  2.00  .75  .50  .90  FRONT

2. DRAW THE TOP, FRONT AND RIGHT SIDE VIEWS.

HOLE DIA. 38 mm.
HOLE CENTERED ON
OBJECT AND PERPENDICULAR
TO THE BACK SURFACE.

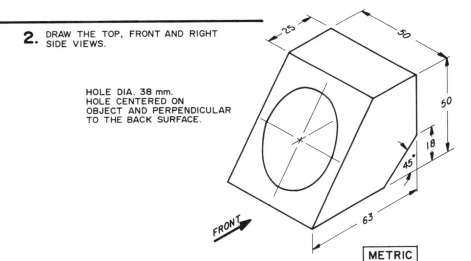

25  50  50  18  45°  63  FRONT

METRIC

# ORTHOGRAPHIC PROJECTION

**I.** DRAW THE FRONT, TOP AND RIGHT SIDE VIEWS.

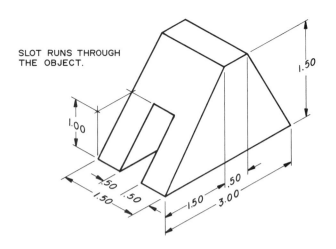

SLOT RUNS THROUGH THE OBJECT.

**2.** DRAW THE TOP, FRONT, AND RIGHT SIDE VIEWS.

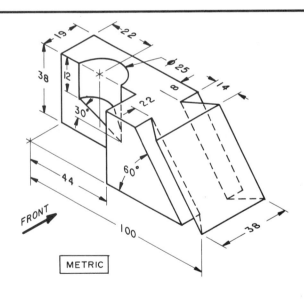

FRONT

METRIC

## ORTHOGRAPHIC PROJECTION

DRAW THE TOP, FRONT AND RIGHT SIDE VIEWS. USE THE LARGEST SCALE POSSIBLE.

4.00
1.50
.75
1.65
.65
.50
.25
.50
2.00
3.00
.75
.75
1.12
Ø.50 THRU
.75 C'BORE
.25 DEEP
3 HOLES
.40
.25
.75
1.00
.50
2.00
3.00

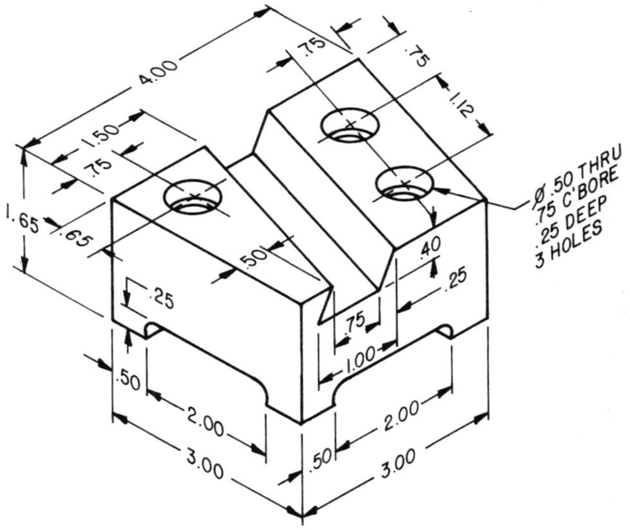

ORTHOGRAPHIC PROJECTION

| ENGINEERING GRAPHICS | NAME<br>CLASS SECTION DATE | GRADE | DRAWING NO. 26 |
|---|---|---|---|

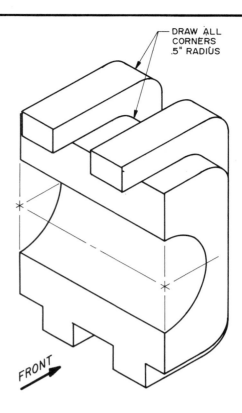

DRAW ALL
CORNERS
.5" RADIUS

FRONT

# ORTHOGRAPHIC PROJECTION

| ENGINEERING GRAPHICS | NAME⌐ | | | GRADE | DRAWING NO. |
|---|---|---|---|---|---|
| | CLASS⌐ | SECTION⌐ | DATE⌐ | | **27** |

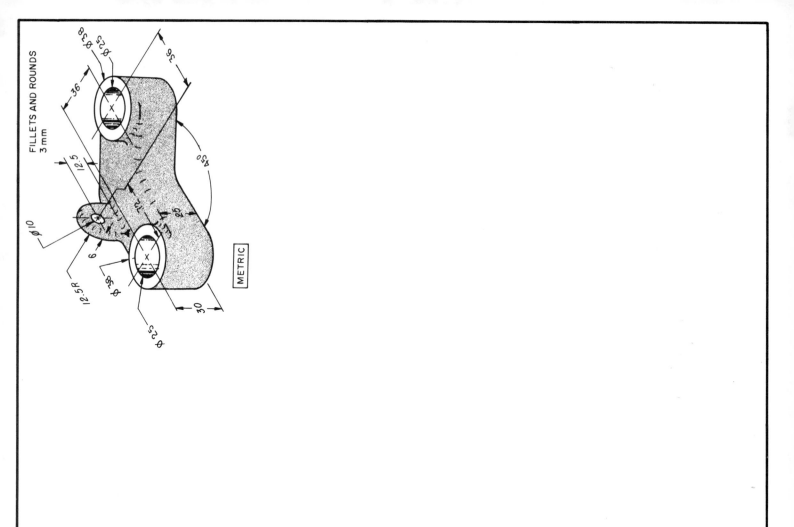

FILLETS AND ROUNDS
3 mm

METRIC

**ORTHOGRAPHIC PROJECTION**

| ENGINEERING GRAPHICS | NAME ⎓ | | | GRADE | DRAWING NO. |
|---|---|---|---|---|---|
| | CLASS ⎓ | SECTION ⎓ | DATE ⎓ | | **28** |

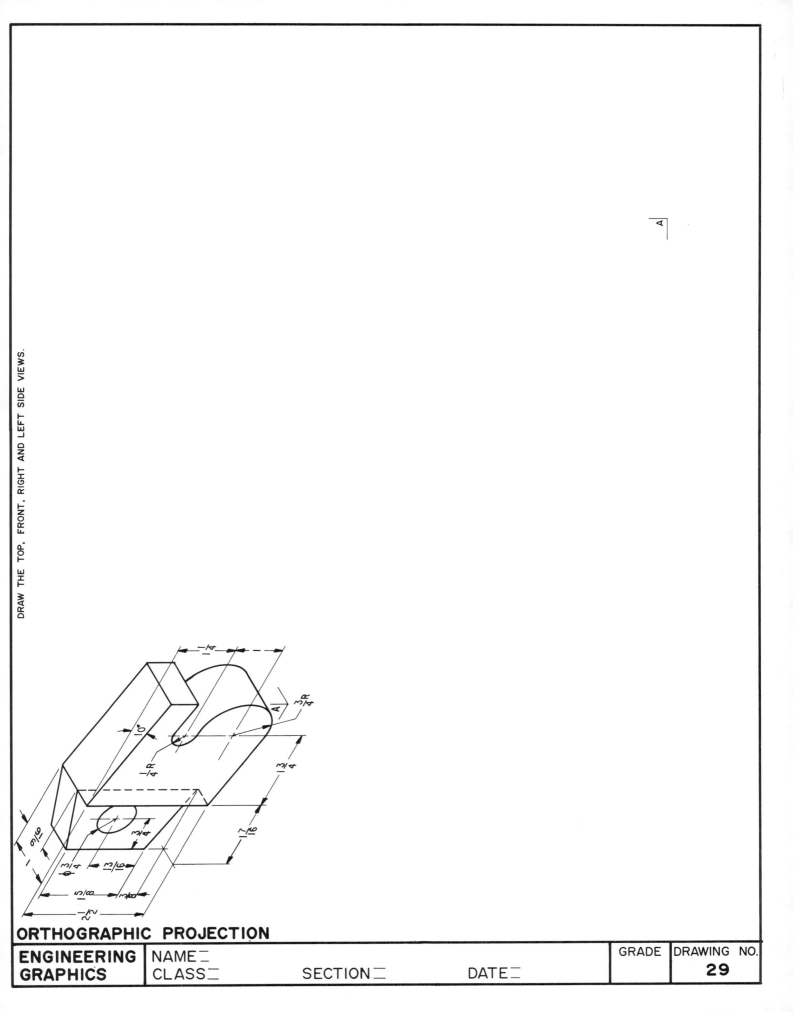

# ORTHOGRAPHIC PROJECTION

| ENGINEERING GRAPHICS | NAME ⊐ | | | GRADE | DRAWING NO. |
|---|---|---|---|---|---|
| | CLASS ⊐ | SECTION ⊐ | DATE ⊐ | | 29 |

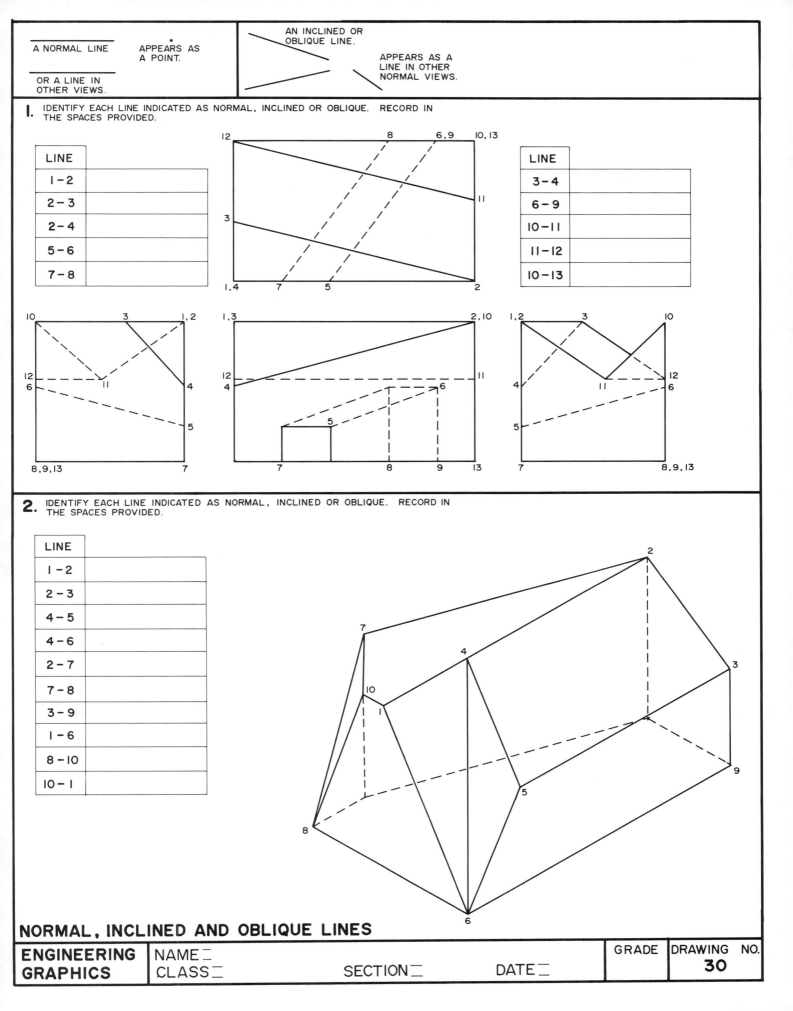

I. IDENTIFY EACH LINE INDICATED AS NORMAL, INCLINED OR OBLIQUE. RECORD IN THE SPACES PROVIDED.

| LINE | |
|------|--|
| 1-2 | |
| 2-3 | |
| 2-4 | |
| 5-6 | |
| 7-8 | |

| LINE | |
|------|--|
| 3-4 | |
| 6-9 | |
| 10-11 | |
| 11-12 | |
| 10-13 | |

2. IDENTIFY EACH LINE INDICATED AS NORMAL, INCLINED OR OBLIQUE. RECORD IN THE SPACES PROVIDED.

| LINE | |
|------|--|
| 1-2 | |
| 2-3 | |
| 4-5 | |
| 4-6 | |
| 2-7 | |
| 7-8 | |
| 3-9 | |
| 1-6 | |
| 8-10 | |
| 10-1 | |

## NORMAL, INCLINED AND OBLIQUE LINES

| ENGINEERING GRAPHICS | NAME= CLASS= | SECTION= DATE= | GRADE | DRAWING NO. 30 |

A PLANE SURFACE

CAN APPEAR
IN ANOTHER
VIEW AS
SHOWN. ⟹

A NORMAL
SURFACE APPEARS
AS AN EDGE.

AN INCLINED
SURFACE APPEARS
AS AN EDGE.

AN INCLINED
SURFACE APPEARS
AS A PLANE.

AN OBLIQUE
SURFACE APPEARS
AS A PLANE.

**I.** IDENTIFY THE SURFACES INDICATED AS NORMAL, INCLINED OR OBLIQUE.
RECORD IN THE SPACES PROVIDED.

| SURFACE | |
|---------|---|
| 1 | |
| 2 | |
| 3 | |
| 4 | |
| 5 | |
| 6 | |
| 7 | |
| 8 | |
| 9 | |
| 10 | |

**2.** IDENTIFY THE SURFACES INDICATED AS NORMAL, INCLINED OR OBLIQUE. RECORD IN
THE SPACES PROVIDED. MAKE A PICTORIAL SKETCH IF YOU HAVE DIFFICULTY.
VISUALIZING THE PRODUCT.

| SURFACE | | SURFACE | |
|---------|---|---------|---|
| 1 | | 4 | |
| 2 | | 5 | |
| 3 | | 6 | |

**NORMAL, INCLINED AND OBLIQUE PLANES**

| ENGINEERING GRAPHICS | NAME ⚊ CLASS ⚊ | SECTION ⚊ | DATE ⚊ | GRADE | DRAWING NO. 31 |
|---|---|---|---|---|---|

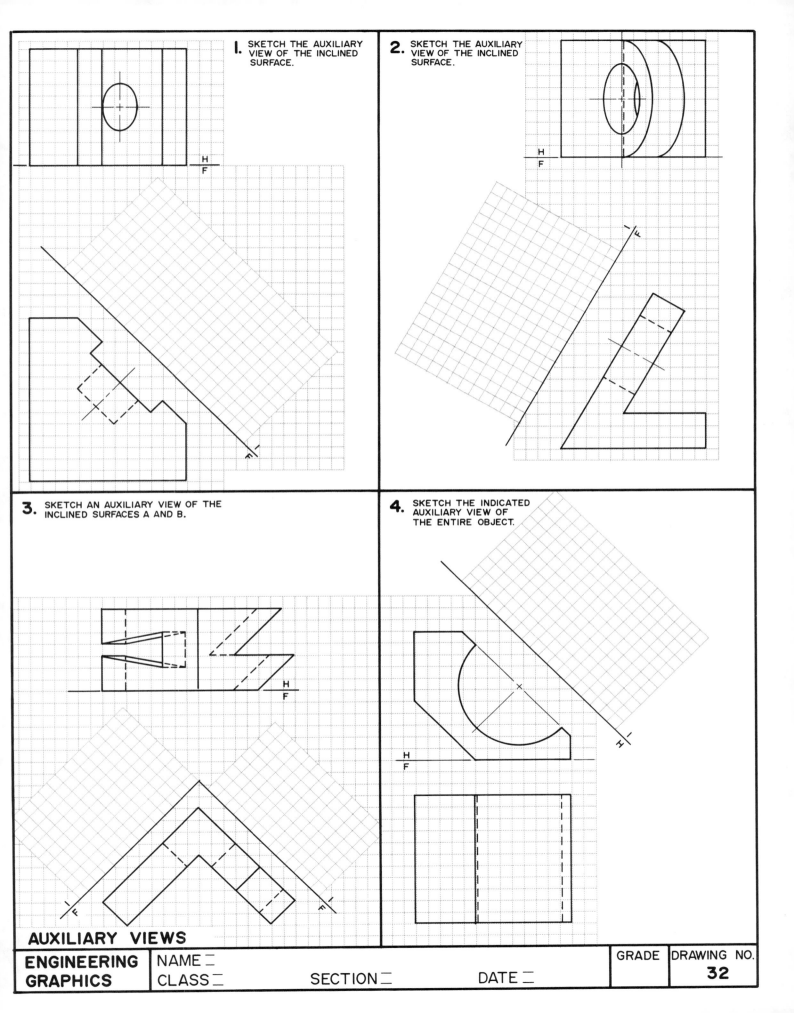

1. SKETCH THE AUXILIARY VIEW OF THE INCLINED SURFACE.

2. SKETCH THE AUXILIARY VIEW OF THE INCLINED SURFACE.

3. SKETCH AN AUXILIARY VIEW OF THE INCLINED SURFACES A AND B.

4. SKETCH THE INDICATED AUXILIARY VIEW OF THE ENTIRE OBJECT.

AUXILIARY VIEWS

| ENGINEERING GRAPHICS | NAME ⁼ | | | GRADE | DRAWING NO. |
|---|---|---|---|---|---|
| | CLASS ⁼ | SECTION ⁼ | DATE ⁼ | | 32 |

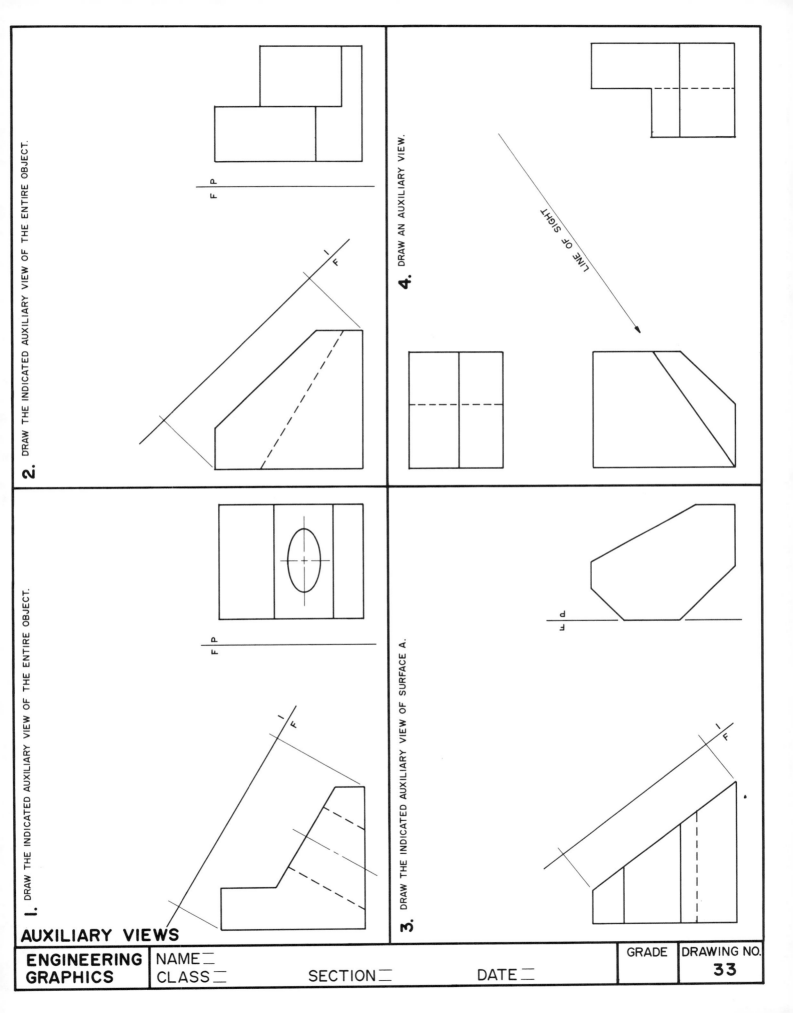

## AUXILIARY VIEWS

**1.** DRAW THE INDICATED AUXILIARY VIEW OF THE ENTIRE OBJECT.

**2.** DRAW THE INDICATED AUXILIARY VIEW OF THE ENTIRE OBJECT.

**3.** DRAW THE INDICATED AUXILIARY VIEW OF SURFACE A.

**4.** DRAW AN AUXILIARY VIEW.

LINE OF SIGHT

| ENGINEERING GRAPHICS | NAME | | | GRADE | DRAWING NO. |
|---|---|---|---|---|---|
| | CLASS | SECTION | DATE | | 33 |

**I.** DRAW THE TWO INDICATED AUXILIARY VIEWS.

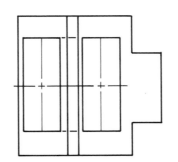

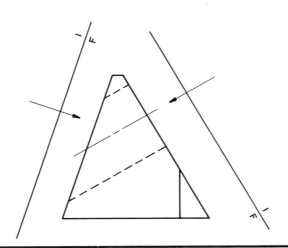

**2.** FIND THE TRUE SIZE OF THE ANGLE BETWEEN PLANES A AND B. MEASURE AND RECORD THE TRUE SIZE OF THE EDGES INDICATED BELOW.

1 – 2 _____

2 – 3 _____

3 – 4 _____

3 – 5 _____

ANGLE _____

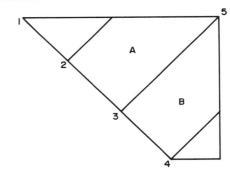

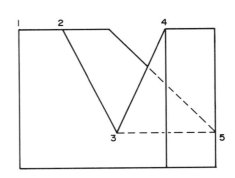

## AUXILIARY VIEWS

| ENGINEERING GRAPHICS | NAME ⊏ | | | GRADE | DRAWING NO. |
| --- | --- | --- | --- | --- | --- |
| | CLASS ⊏ | SECTION ⊏ | DATE ⊏ | | **34** |

**I.** FIND THE TRUE SIZE OF SURFACE 1-2-3 BY MAKING A SECONDARY AUXILIARY. MEASURE AND RECORD THE TRUE SIZE OF THE ANGLE AT EACH CORNER.

ANGLES

CORNER 1 _____

CORNER 2 _____

CORNER 3 _____

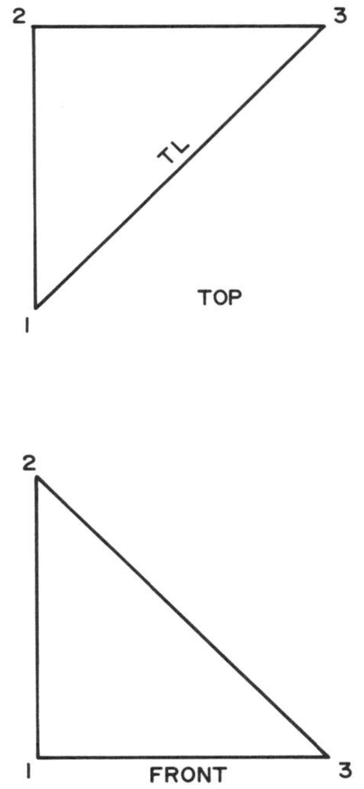

TOP

FRONT

**2.** FIND THE TRUE SIZE OF THE OBLIQUE PLANE. RECORD THE TRUE SIZE OF THE ANGLES AT CORNERS 1, 6, 7, AND 8.

ANGLES

CORNER 1 _____

CORNER 6 _____

CORNER 7 _____

CORNER 8 _____

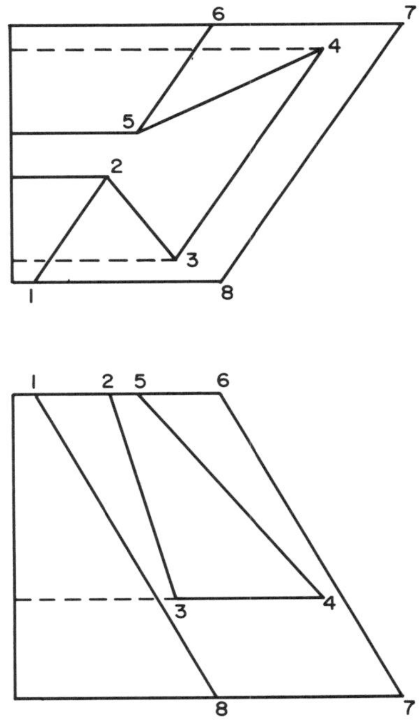

# SECONDARY AUXILIARY VIEWS

| ENGINEERING GRAPHICS | NAME⁼ | | | GRADE | DRAWING NO. |
| --- | --- | --- | --- | --- | --- |
| | CLASS⁼ | SECTION⁼ | DATE⁼ | | 35 |

DRAW THE MISSING VIEWS IN EACH PROBLEM. LOCATE THE POINTS
ON THE LINES IN EACH VIEW. IDENTIFY EACH TRUE LENGTH LINE, TL.

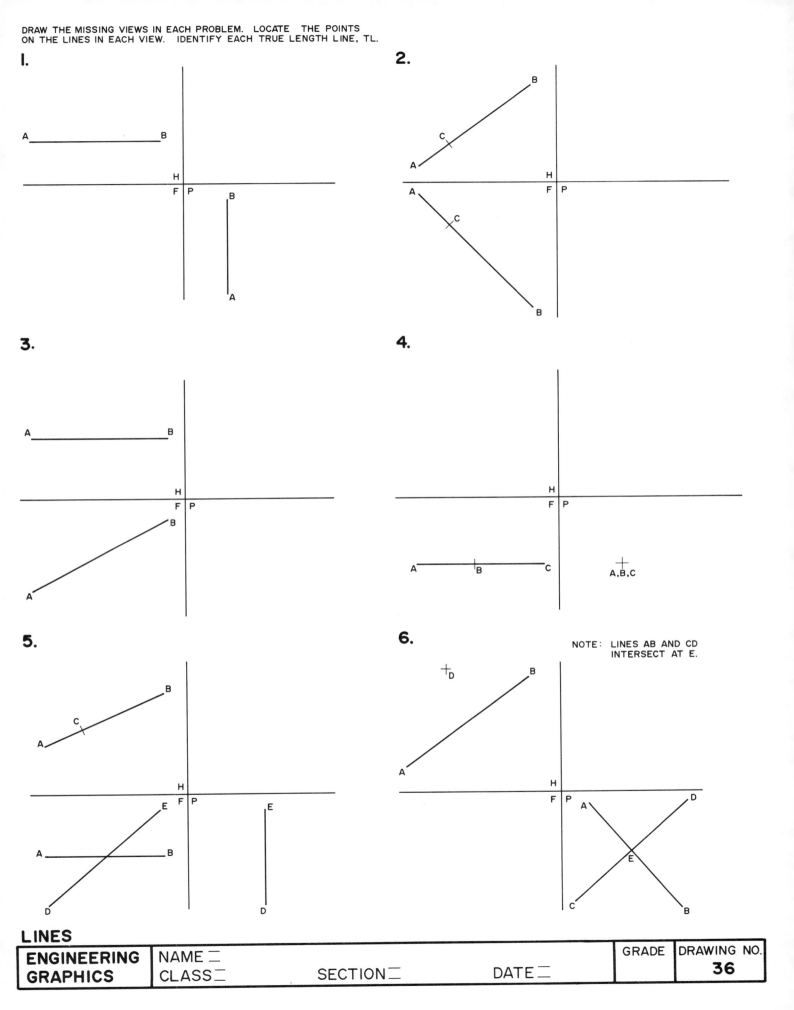

**1.**

**2.**

**3.**

**4.**

**5.**

**6.**

NOTE: LINES AB AND CD
INTERSECT AT E.

**LINES**

| ENGINEERING GRAPHICS | NAME ⎓  CLASS ⎓         SECTION ⎓         DATE ⎓ | GRADE | DRAWING NO. 36 |

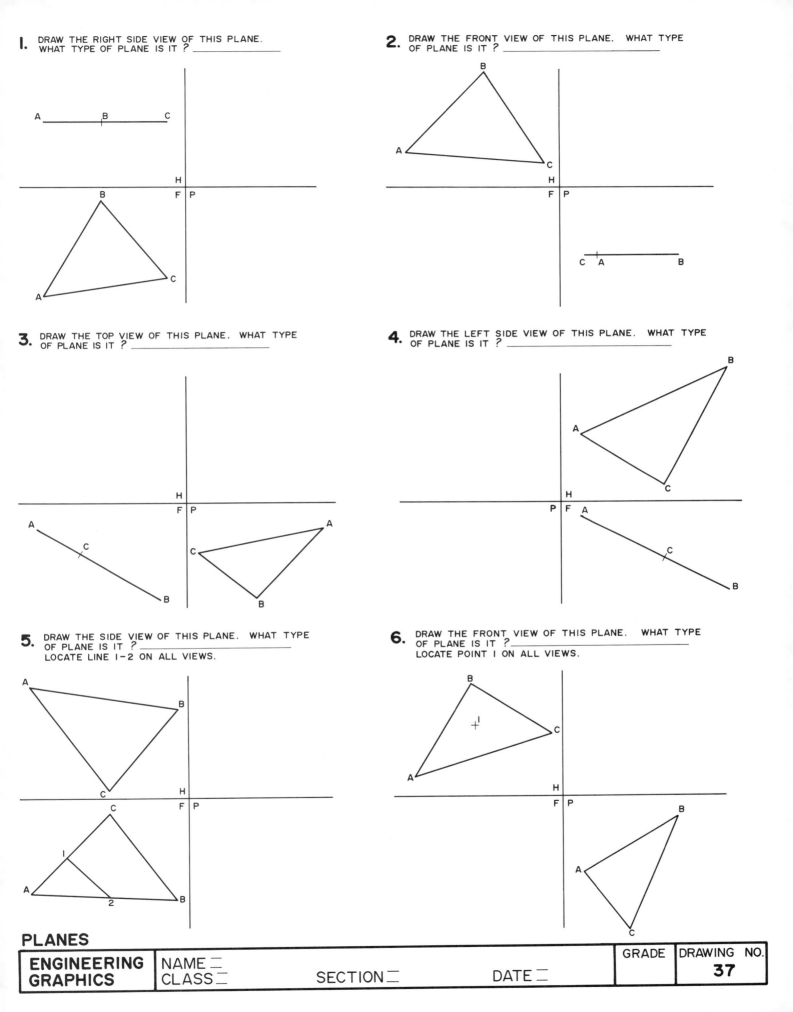

**1.** DRAW THE RIGHT SIDE VIEW OF THIS PLANE. WHAT TYPE OF PLANE IS IT ? _____

**2.** DRAW THE FRONT VIEW OF THIS PLANE. WHAT TYPE OF PLANE IS IT ? _____

**3.** DRAW THE TOP VIEW OF THIS PLANE. WHAT TYPE OF PLANE IS IT ? _____

**4.** DRAW THE LEFT SIDE VIEW OF THIS PLANE. WHAT TYPE OF PLANE IS IT ? _____

**5.** DRAW THE SIDE VIEW OF THIS PLANE. WHAT TYPE OF PLANE IS IT ? _____
LOCATE LINE 1-2 ON ALL VIEWS.

**6.** DRAW THE FRONT VIEW OF THIS PLANE. WHAT TYPE OF PLANE IS IT ? _____
LOCATE POINT 1 ON ALL VIEWS.

**PLANES**

| ENGINEERING GRAPHICS | NAME ⎓ CLASS ⎓ | SECTION ⎓ | DATE ⎓ | GRADE | DRAWING NO. 37 |

**1.** DRAW A LINE FROM POINT I THAT INTERSECTS LINE AB AT POINT C. COMPLETE THE FRONT AND SIDE VIEWS.

**2.** DO LINES AB AND CD INTERSECT? _____

**3.** DO LINES AB AND CD INTERSECT? _____

**4.** DETERMINE THE VISIBILITY AND COMPLETE EACH VIEW.

**5.** DETERMINE THE VISIBILITY AND COMPLETE EACH VIEW.

**6.** LOCATE THE POINT WHERE LINE 1-2 PIERCES PLANE ABC. DETERMINE VISIBILITY AND COMPLETE THE TOP, FRONT AND RIGHT SIDE VIEWS.

# INTERSECTION, VISIBILITY AND PIERCING POINTS

| ENGINEERING GRAPHICS | NAME ⎓ CLASS ⎓  SECTION ⎓  DATE ⎓ | GRADE | DRAWING NO. 38 |

**1.** ESTABLISH A FRONTAL PLANE THAT PASSES THROUGH POINT A. DRAW A LINE ON A 45° ANGLE THAT PIERCES THE PLANE AT POINT A. COMPLETE THE TOP, FRONT, AND SIDE VIEWS.

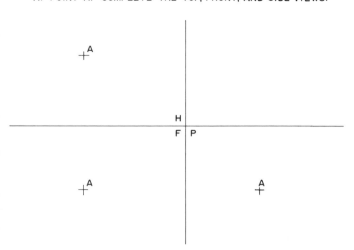

**2.** FIND THE POINT WHERE LINE 1-2 PIERCES PLANE ABC. COMPLETE TOP, FRONT AND LEFT SIDE VIEWS.

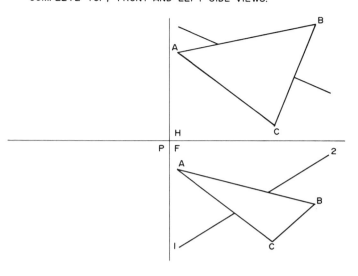

**3.** FIND THE POINT WHERE LINE 1-2 PIERCES PLANE ABCD. COMPLETE THE TOP, FRONT AND SIDE VIEWS.

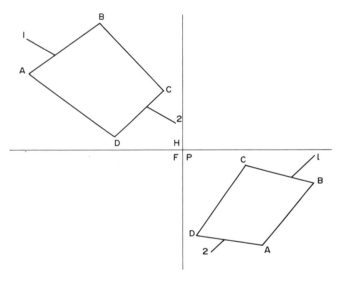

**4.** DRAW A LINE THAT PIERCES PLANE ABC AT POINT 1. COMPLETE THE TOP, FRONT AND SIDE VIEWS.

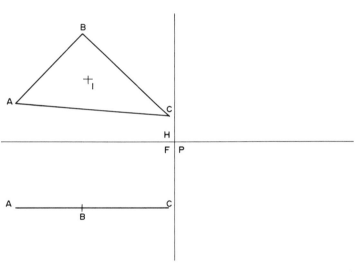

**5.** DRAW A LINE THAT PIERCES PLANE ABC AT POINT 1. COMPLETE THE TOP, FRONT AND SIDE VIEWS.

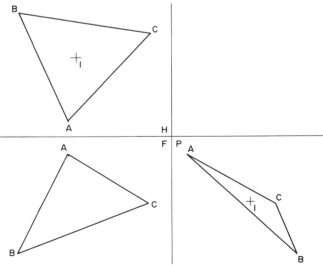

**6.** DETERMINE THE VISIBILITY BETWEEN LINES 1-2, 3-4 AND PLANE ABC. COMPLETE THE TOP, FRONT AND SIDE VIEWS.

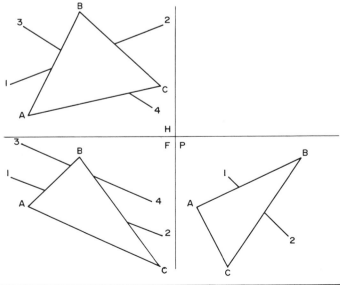

# PIERCING POINTS AND VISIBILITY

| ENGINEERING GRAPHICS | NAME = CLASS = | SECTION = | DATE = | GRADE | DRAWING NO. 39 |

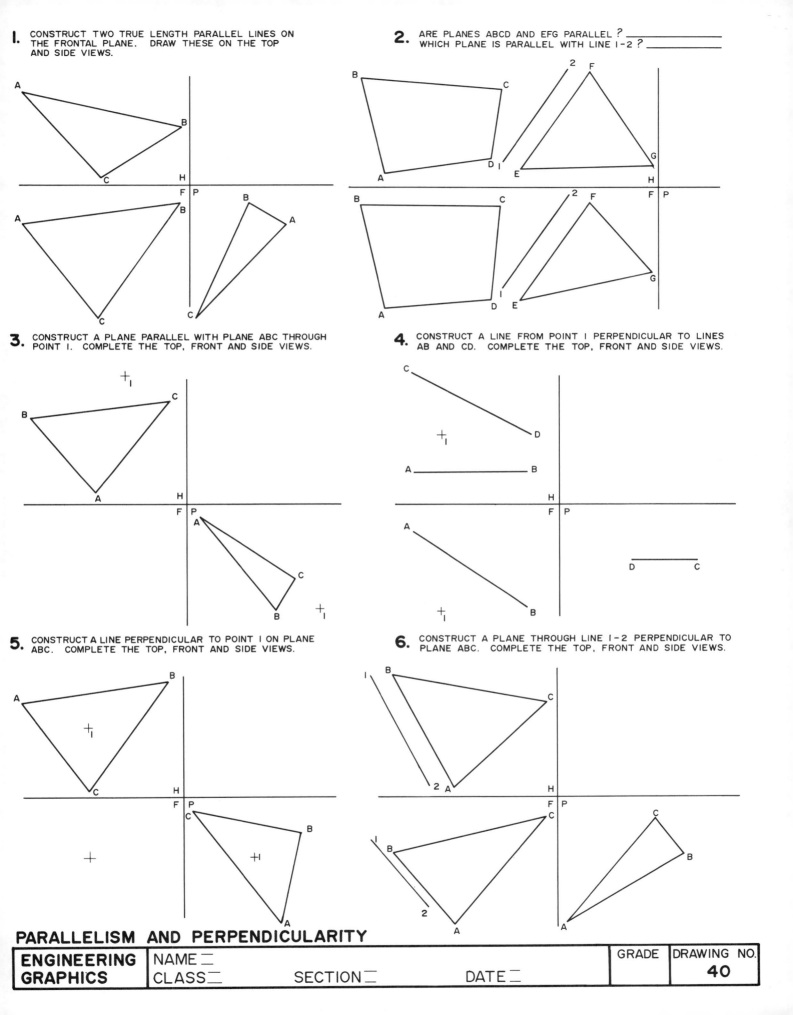

**1.** CONSTRUCT TWO TRUE LENGTH PARALLEL LINES ON THE FRONTAL PLANE. DRAW THESE ON THE TOP AND SIDE VIEWS.

**2.** ARE PLANES ABCD AND EFG PARALLEL ? _____
WHICH PLANE IS PARALLEL WITH LINE 1-2 ? _____

**3.** CONSTRUCT A PLANE PARALLEL WITH PLANE ABC THROUGH POINT 1. COMPLETE THE TOP, FRONT AND SIDE VIEWS.

**4.** CONSTRUCT A LINE FROM POINT 1 PERPENDICULAR TO LINES AB AND CD. COMPLETE THE TOP, FRONT AND SIDE VIEWS.

**5.** CONSTRUCT A LINE PERPENDICULAR TO POINT 1 ON PLANE ABC. COMPLETE THE TOP, FRONT AND SIDE VIEWS.

**6.** CONSTRUCT A PLANE THROUGH LINE 1-2 PERPENDICULAR TO PLANE ABC. COMPLETE THE TOP, FRONT AND SIDE VIEWS.

# PARALLELISM AND PERPENDICULARITY

| ENGINEERING GRAPHICS | NAME ⎓ | | | GRADE | DRAWING NO. |
|---|---|---|---|---|---|
| | CLASS ⎓ | SECTION ⎓ | DATE ⎓ | | 40 |

**1.** FIND THE TRUE LENGTH OF LINE 1-2 BY TAKING AUXILIARY VIEWS OF THE TOP AND SIDE VIEWS. MEASURE THE LENGTH IN INCHES AND MILLIMETERS AND RECORD ON THE TRUE LENGTH LINES.

**2.** FIND THE TRUE LENGTH OF LINES 1-2 AND 3-4. MEASURE THE LENGTH IN INCHES AND RECORD ON THE TRUE LENGTH LINES.

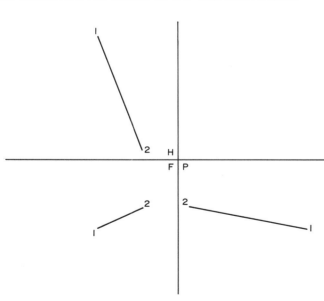
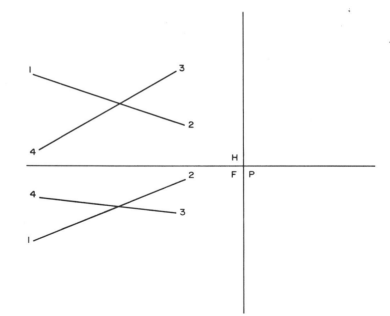

**3.** FIND THE TRUE LENGTH OF EACH PART OF THE SLING BELOW. RECORD THE TRUE SIZES IN THE SPACES BELOW.

|  | INCHES | MILLIMETERS |
|---|---|---|
| A B | _____ | _____ |
| A C | _____ | _____ |
| A D | _____ | _____ |
| A E | _____ | _____ |

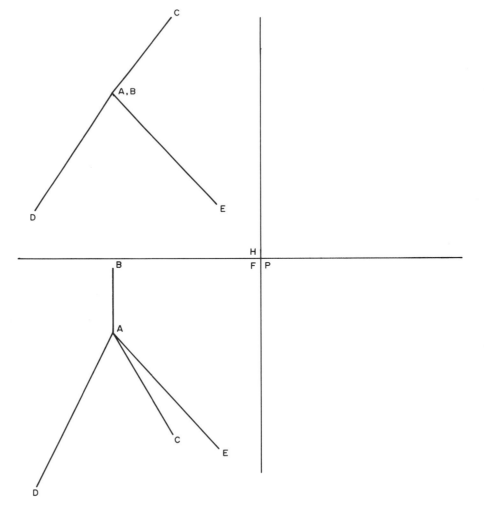

**TRUE LENGTH LINES**

| ENGINEERING GRAPHICS | NAME⎯ CLASS⎯ | SECTION⎯ | DATE⎯ | GRADE | DRAWING NO. 41 |

**1.** FIND THE TRUE SIZE OF THE ANGLE LINE 1-2 MAKES WITH THE HORIZONTAL, FRONTAL AND PROFILE PLANES. DIMENSION THE ANGLE WHERE IT APPEARS TRUE SIZE.

**2.** FIND THE TRUE SIZE OF THE ANGLE EDGE 1-2 MAKES WITH THE FRONTAL PLANE AND 1-3 MAKES WITH THE HORIZONTAL PLANE. DIMENSION THE ANGLE WHERE IT APPEARS TRUE SIZE.

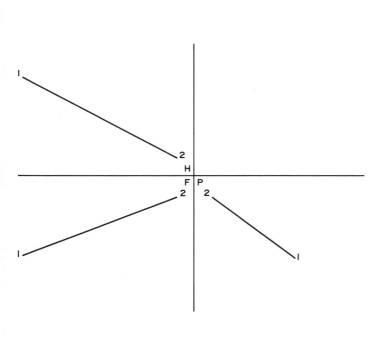

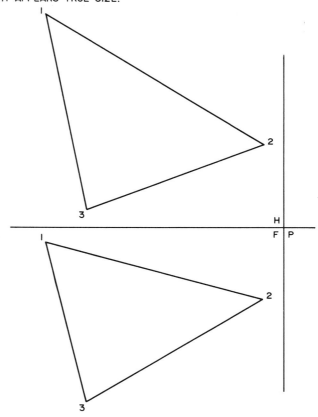

**3.** FIND THE TRUE SIZE OF THE ANGLE BETWEEN THE PLANES BELOW. DIMENSION THE ANGLE WHERE IT APPEARS TRUE SIZE.

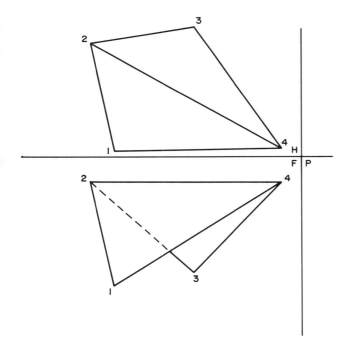

## TRUE SIZE OF ANGLES

| ENGINEERING GRAPHICS | NAME ⎓ | | | GRADE | DRAWING NO. |
|---|---|---|---|---|---|
| | CLASS ⎓ | SECTION ⎓ | DATE ⎓ | | **42** |

**1.** FIND THE POINT WHERE THE CABLE, I-2, PIERCES THE WALL, ABC USING THE AUXILIARY VIEW METHOD. PROJECT THE AUXILIARY VIEW OFF OF THE TOP VIEW. DETERMINE VISIBILITY.

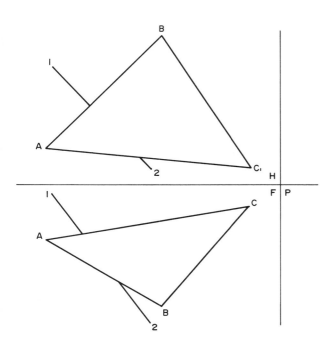

**2.** FIND THE POINT WHERE LINE I-2 PIERCES PLANE ABC USING THE PROJECTION METHOD. DETERMINE VISIBILITY.

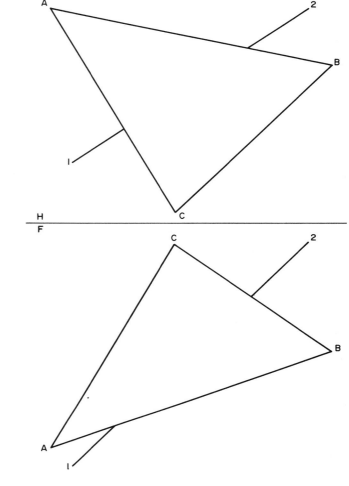

**3.** WHICH LINES PIERCE PLANE ABC ? RECORD YOUR FINDINGS IN THE CHART PROVIDED. LOCATE THE PIERCING POINTS OF THOSE LINES THAT DO PIERCE THE PLANE. DETERMINE VISIBILITY.

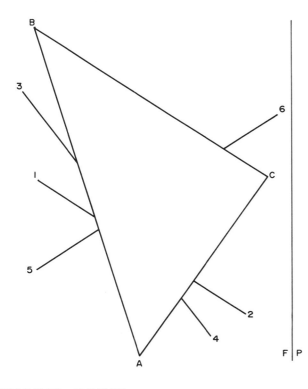

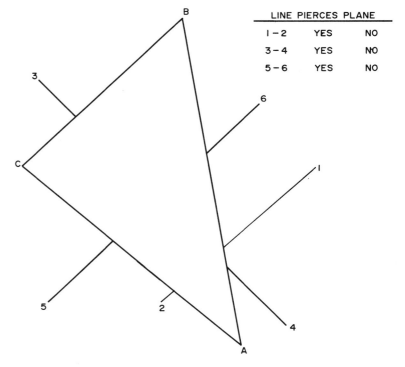

| LINE PIERCES PLANE | | |
|---|---|---|
| I - 2 | YES | NO |
| 3 - 4 | YES | NO |
| 5 - 6 | YES | NO |

**PIERCING POINT**

| ENGINEERING GRAPHICS | NAME⁼ CLASS⁼    SECTION⁼        DATE⁼ | | GRADE | DRAWING NO. 43 |
|---|---|---|---|---|

**I.** FIND THE LINE OF INTERSECTION OF THE TWO PLANES. TAKE THE AUXILIARY VIEW OFF THE FRONT VIEW. PROJECT THE INTERSECTION TO ALL VIEWS. DETERMINE VISIBILITY.

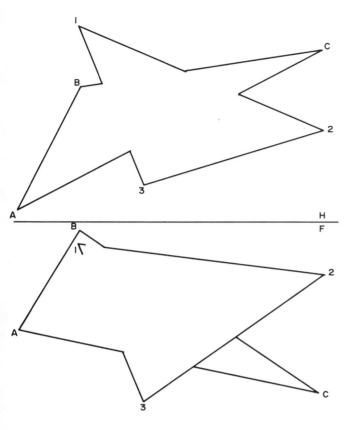

**2.** FIND THE LINE OF INTERSECTION BETWEEN THESE TWO PLANES. TAKE THE AUXILIARY VIEW OFF THE TOP VIEW. PROJECT THE INTERSECTION TO ALL VIEWS. DETERMINE VISIBILITY.

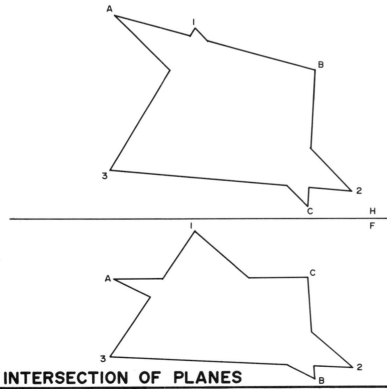

# INTERSECTION OF PLANES

| ENGINEERING GRAPHICS | NAME ⎓  CLASS ⎓    SECTION ⎓    DATE ⎓ | GRADE | DRAWING NO. 44 |

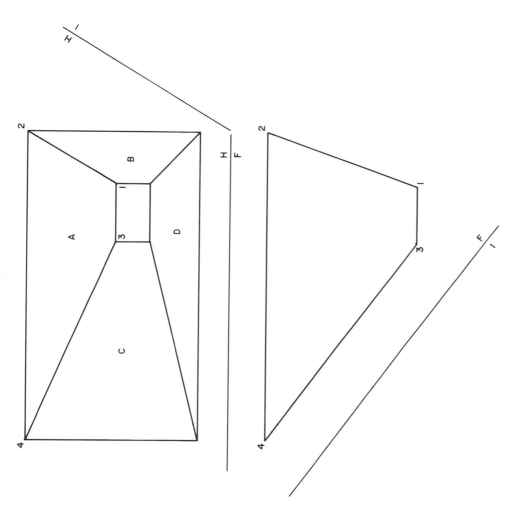

FIND THE TRUE SIZE OF THE ANGLE BETWEEN PLANES A AND B BY FINDING EDGE I-2 IN POINT VIEW OFF THE TOP VIEW AND THE ANGLE BETWEEN PLANES A AND C BY FINDING EDGE 3-4 IN POINT VIEW OFF THE FRONT VIEW. RECORD THE SIZE OF THE ANGLE WHERE IT APPEARS TRUE SIZE.

# ANGLE BETWEEN OBLIQUE PLANES

| **ENGINEERING** | NAME⁼ | | | GRADE | DRAWING NO. |
|---|---|---|---|---|---|
| **GRAPHICS** | CLASS⁼ | SECTION⁼ | DATE⁼ | | **45** |

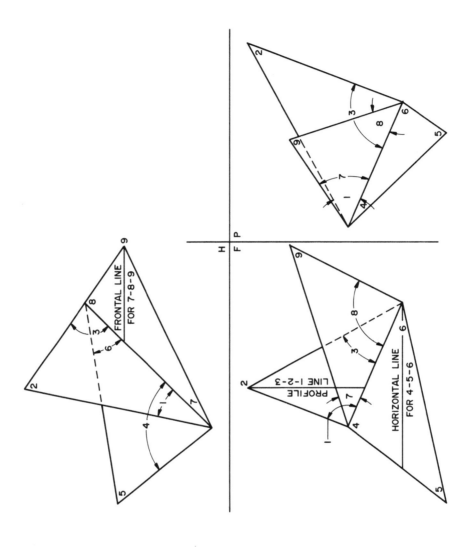

**TRUE SIZE OF OBLIQUE PLANES**

| ENGINEERING GRAPHICS | NAME ‗ | | | GRADE | DRAWING NO. |
|---|---|---|---|---|---|
| | CLASS ‗ | SECTION ‗ | DATE ‗ | | **46** |

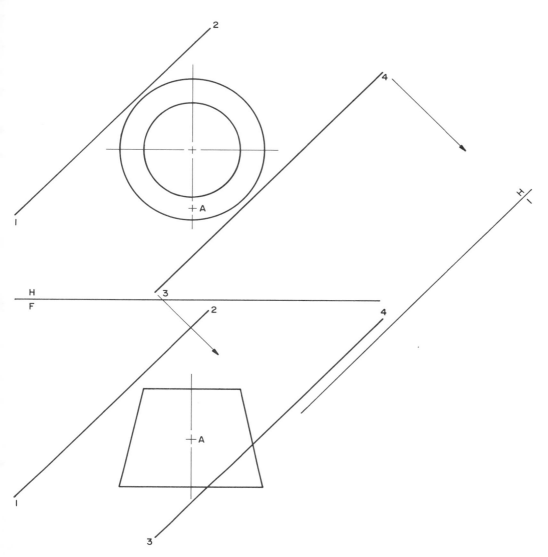

1. FIND THE SHORTEST DISTANCE BETWEEN POINT
   A AND LINE 1-2. PROJECT IT TO ALL VIEWS. FIND
   THE SHORTEST DISTANCE BETWEEN LINES 1-2
   AND 3-4. MEASURE IT IN INCHES AND RECORD
   ON THE DRAWING.  CONSTRUCT A LINE
   REPRESENTING THE SHORTEST DISTANCE
   BETWEEN 1-2 AND 3-4.

2. FIND THE SHORTEST DISTANCE BETWEEN THESE
   SKEW LINES.  PROJECT TO ALL VIEWS.  FIND
   THE POINT VIEW OF LINE 1-2.

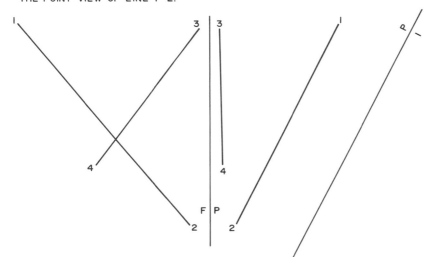

# SHORTEST DISTANCE BETWEEN POINTS AND LINES

| ENGINEERING GRAPHICS | NAME ‗ CLASS ‗  SECTION ‗  DATE ‗ | GRADE | DRAWING NO. 47 |
| --- | --- | --- | --- |

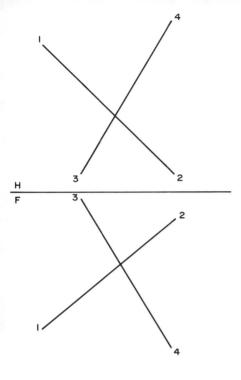

**1.** FIND THE SHORTEST LINE HAVING A SLOPE OF 30° WITH THE HORIZONTAL THAT CAN CONNECT THESE SKEW LINES. USE LINE 1-2 TO FORM THE NEEDED PLANE. PROJECT THE LINE TO ALL VIEWS.

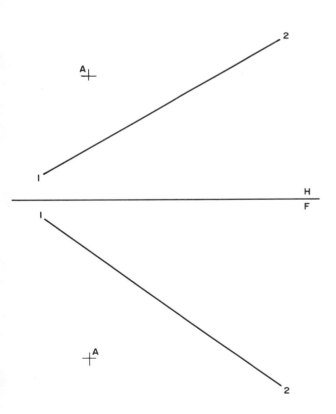

**2.** CONSTRUCT A LINE FROM POINT A TO LINE 1-2 AT AN ANGLE OF 60° TO LINE 1-2. PROJECT IT TO ALL VIEWS.

# CONSTRUCTING SLOPED LINES TO KNOWN LINES

| ENGINEERING GRAPHICS | NAME ⎓ | | | GRADE | DRAWING NO. |
|---|---|---|---|---|---|
| | CLASS ⎓ | SECTION ⎓ | DATE ⎓ | | **48** |

1. FIND THE TRUE SIZE OF THE ANGLE BETWEEN THE LINE AND PLANE. FIND THE PIERCING POINT AND NOTE VISIBILITY IN ALL VIEWS. PROJECT THE AUXILIARY VIEWS OFF THE TOP VIEW.

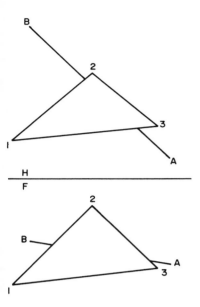

2. FIND THE TRUE SIZE OF THE ANGLE BETWEEN THE LINE AND THE PLANE. FIND THE PIERCING POINT AND NOTE IT IN ALL VIEWS. PROJECT THE AUXILIARY VIEWS OFF THE FRONT VIEW.

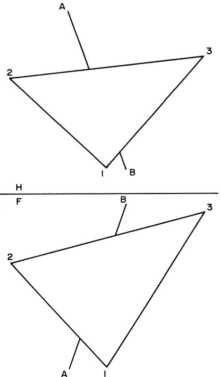

# TRUE SIZE OF ANGLE BETWEEN A LINE AND AN OBLIQUE PLANE

| ENGINEERING GRAPHICS | NAME ⹂ CLASS ⹂ | SECTION ⹂ | DATE ⹂ | GRADE | DRAWING NO. 49 |

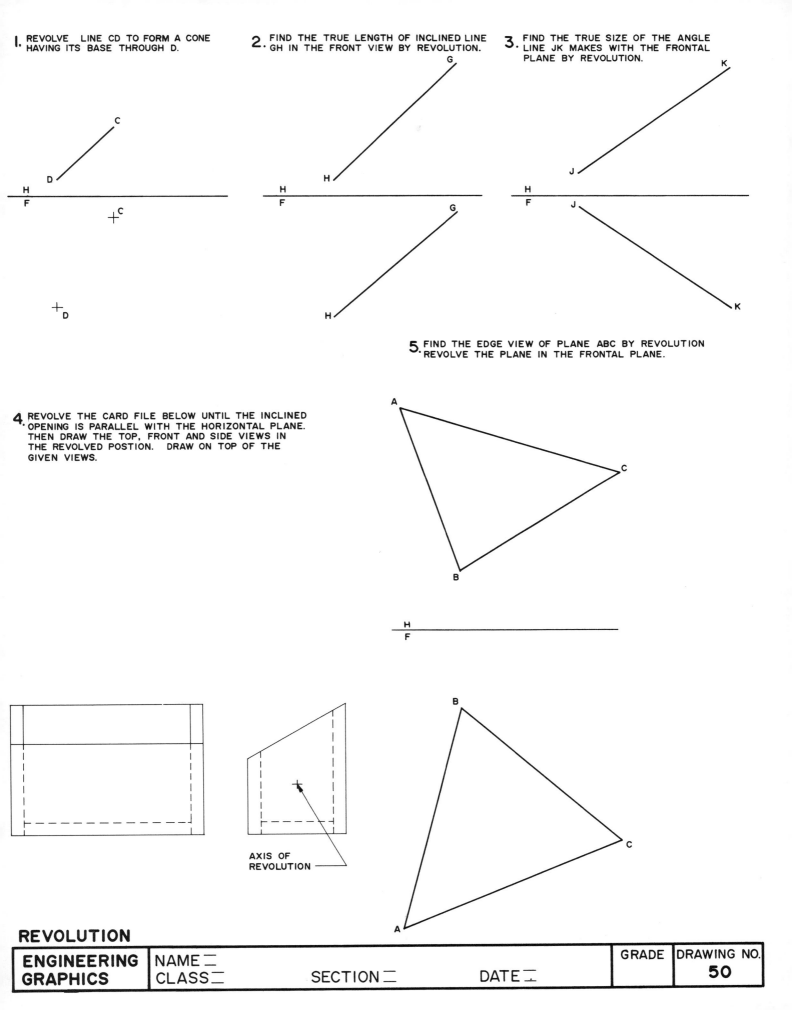

**1.** REVOLVE LINE CD TO FORM A CONE HAVING ITS BASE THROUGH D.

C
D
H
F
+C
+D

**2.** FIND THE TRUE LENGTH OF INCLINED LINE GH IN THE FRONT VIEW BY REVOLUTION.

G
H
H
F
G
H

**3.** FIND THE TRUE SIZE OF THE ANGLE LINE JK MAKES WITH THE FRONTAL PLANE BY REVOLUTION.

K
J
H
F
J
K

**5.** FIND THE EDGE VIEW OF PLANE ABC BY REVOLUTION. REVOLVE THE PLANE IN THE FRONTAL PLANE.

A
C
B
H
F

**4.** REVOLVE THE CARD FILE BELOW UNTIL THE INCLINED OPENING IS PARALLEL WITH THE HORIZONTAL PLANE. THEN DRAW THE TOP, FRONT AND SIDE VIEWS IN THE REVOLVED POSTION. DRAW ON TOP OF THE GIVEN VIEWS.

AXIS OF REVOLUTION

B
C
A

## REVOLUTION

| ENGINEERING GRAPHICS | NAME CLASS | SECTION | DATE | GRADE | DRAWING NO. |
|---|---|---|---|---|---|
| | | | | | **50** |

**1.** FIND THE TRUE SIZE OF THE ANGLE BETWEEN THESE PLANES BY REVOLUTION. RECORD THE SIZE ON THE DRAWING.

**2.** FIND THE TRUE SIZE OF THIS OBLIQUE PLANE USING AUXILIARY VIEWS AND REVOLUTION. TAKE THE AUXILIARY VIEW OFF OF THE TOP VIEW. LIGHTLY SHADE THE TRUE SIZE OF THE PLANE.

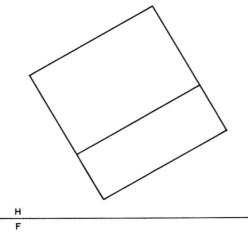

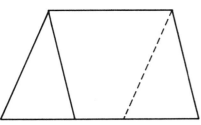

H
F

H
F

**3.** FIND THE TRUE SIZE OF THE ANGLE BETWEEN THESE OBLIQUE PLANES USING A PRIMARY AUXILIARY AND REVOLUTION. RECORD THE SIZE OF THE ANGLE ON THE DRAWING.

H F

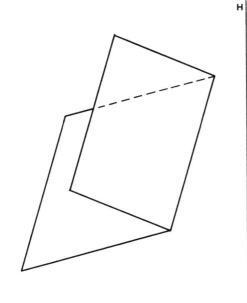

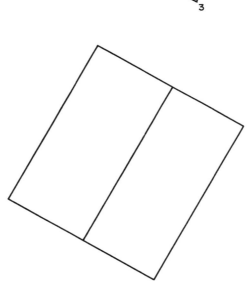

TRUE SIZE OF ANGLE BETWEEN PLANES

| ENGINEERING GRAPHICS | NAME ⎓ | | | GRADE | DRAWING NO. |
|---|---|---|---|---|---|
| | CLASS ⎓ | SECTION ⎓ | DATE ⎓ | | **51** |

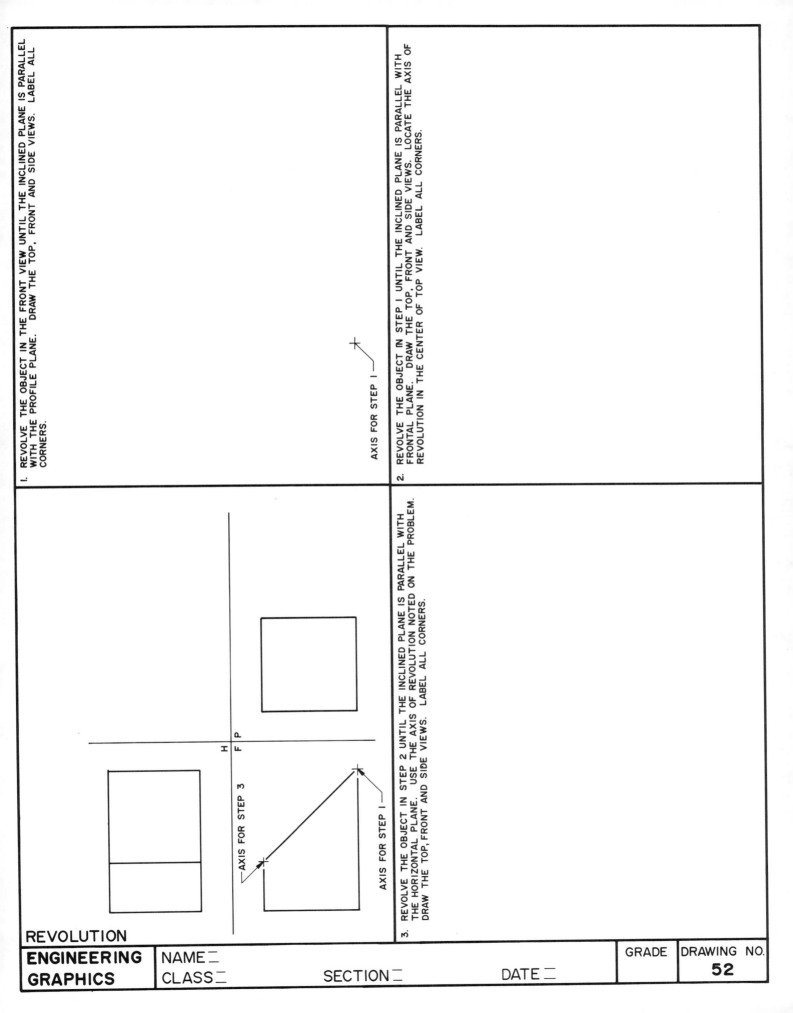

1. REVOLVE THE OBJECT IN THE FRONT VIEW UNTIL THE INCLINED PLANE IS PARALLEL WITH THE PROFILE PLANE. DRAW THE TOP, FRONT AND SIDE VIEWS. LABEL ALL CORNERS.

AXIS FOR STEP 1

2. REVOLVE THE OBJECT IN STEP 1 UNTIL THE INCLINED PLANE IS PARALLEL WITH FRONTAL PLANE. DRAW THE TOP, FRONT AND SIDE VIEWS. LOCATE THE AXIS OF REVOLUTION IN THE CENTER OF TOP VIEW. LABEL ALL CORNERS.

H P
F P

AXIS FOR STEP 3

AXIS FOR STEP 1

3. REVOLVE THE OBJECT IN STEP 2 UNTIL THE INCLINED PLANE IS PARALLEL WITH THE HORIZONTAL PLANE. USE THE AXIS OF REVOLUTION NOTED ON THE PROBLEM. DRAW THE TOP, FRONT AND SIDE VIEWS. LABEL ALL CORNERS.

REVOLUTION

**ENGINEERING**
**GRAPHICS**

NAME
CLASS          SECTION          DATE

GRADE | DRAWING NO.
**52**

MAKE CAVALIER OBLIQUE DRAWINGS OF THESE PRODUCTS.

**1.** DRAW FULL SIZE.
USE A 45° RECEDING AXIS.

4

2

1 R    ½ R

½ R

2½

2

½ R

A

½

PIPE HANGER

+A

**2.** DRAW FULL SIZE.
USE A 30° RECEDING AXIS.

METRIC

+B

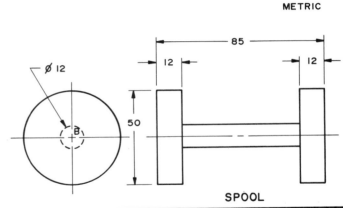

85

12    12

Ø 12

B

50

SPOOL

# OBLIQUE DRAWING

| ENGINEERING<br>GRAPHICS | NAME ⊐<br>CLASS ⊐ | SECTION ⊐ | DATE ⊐ | GRADE | DRAWING NO.<br>**53** |

MAKE CABINET OBLIQUE DRAWINGS OF THESE PRODUCTS.

**1.** DRAW FULL SIZE.
USE A 45° RECEDING AXIS.

Ø .75 –
2 HOLES

$1\frac{1}{16}$

$3\frac{1}{2}$

$2\frac{3}{4}$

A

$1\frac{1}{16}$

3

$\frac{1}{2}$

A

PARTIAL FRONT
VIEW

45° CHAMFER
SECTION VIEW

BEARING CAP

+A

**2.** DRAW FULL SIZE.
USE A 30° RECEDING AXIS.

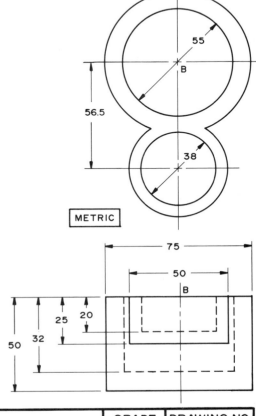

55

B

56.5

38

METRIC

+B

75

50

B

25   20

32

50

OBLIQUE DRAWING

| ENGINEERING GRAPHICS | NAME ⸗ CLASS ⸗ | SECTION ⸗ | DATE ⸗ | GRADE | DRAWING NO. 54 |

DRAW A CABINET DETAIL OF THESE
KITCHEN CABINETS. USE THE
SCALE 1" = 1'-0".

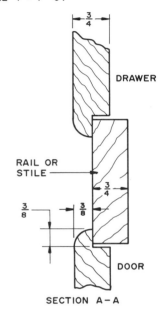

SECTION A-A

NOTE: BASE CABINET 24" DEEP.
WALL CABINET 12" DEEP.
RAILS AND STILES 2" WIDE.
DRAWER FRONT 4" WIDE.

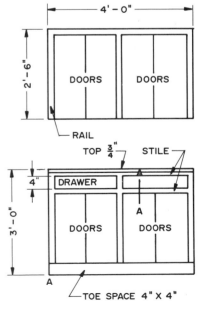

**OBLIQUE DRAWING**

| ENGINEERING GRAPHICS | NAME ⹀ CLASS ⹀ | SECTION ⹀ | DATE ⹀ | GRADE | DRAWING NO. 55 |
|---|---|---|---|---|---|

1. MAKE AN ISOMETRIC DRAWING OF
   THIS PRODUCT, SCALE FULL SIZE.
   USE POINT A AS THE LOCATION
   OF THE CENTER LINE.

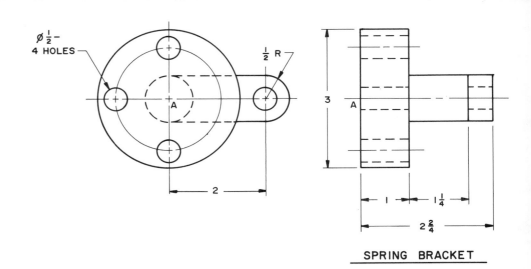

SPRING BRACKET

2. MAKE AN ISOMETRIC DRAWING
   OF THIS PRODUCT. SCALE
   FULL SIZE. USE POINT A TO
   LOCATE IT ON THE PAGE.

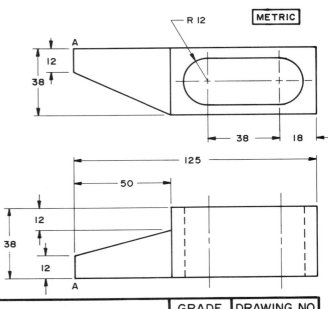

METRIC

# ISOMETRIC DRAWING

| **ENGINEERING** | NAME⸗ | | | GRADE | DRAWING NO. |
| **GRAPHICS** | CLASS⸗ | SECTION⸗ | DATE⸗ | | **56** |

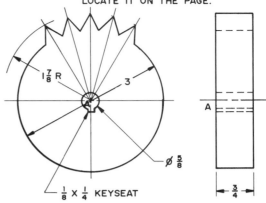

1. DRAW FULL SIZE. THE TEETH ARE ON 15° CENTERS. USE POINT A TO LOCATE IT ON THE PAGE.

$1\frac{7}{8}$ R

3

$\emptyset \frac{5}{8}$

$\frac{1}{8} \times \frac{1}{4}$ KEYSEAT

A

$\frac{3}{4}$

A
+

**CONTROL GEAR**

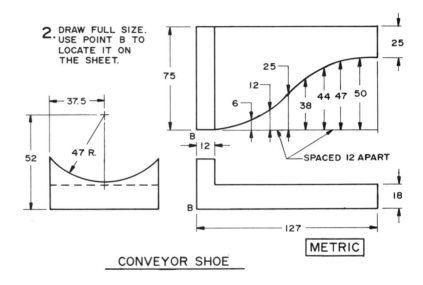

2. DRAW FULL SIZE. USE POINT B TO LOCATE IT ON THE SHEET.

37.5

47 R.

52

75

25

12

6

25

38

44  47  50

12

SPACED 12 APART

B

18

127

METRIC

**CONVEYOR SHOE**

B
+

# ISOMETRIC DRAWING

MAKE ISOMETRIC DRAWINGS OF THE ASSIGNED PROBLEM.   ONLY ONE SOLUTION WILL FIT ON THIS PAGE.

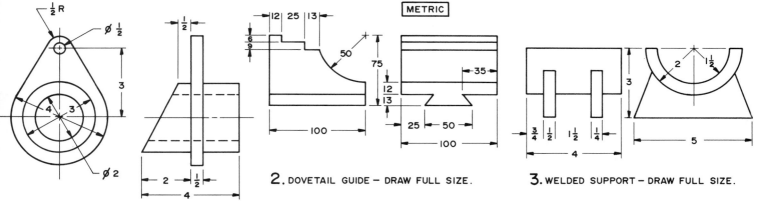

1. SHAFT HANGER — DRAW FULL SIZE.

2. DOVETAIL GUIDE — DRAW FULL SIZE.

3. WELDED SUPPORT — DRAW FULL SIZE.

METRIC

## ISOMETRIC DRAWING

| ENGINEERING GRAPHICS | NAME ⎓ CLASS ⎓ | SECTION ⎓ | DATE ⎓ | GRADE | DRAWING NO. 58 |

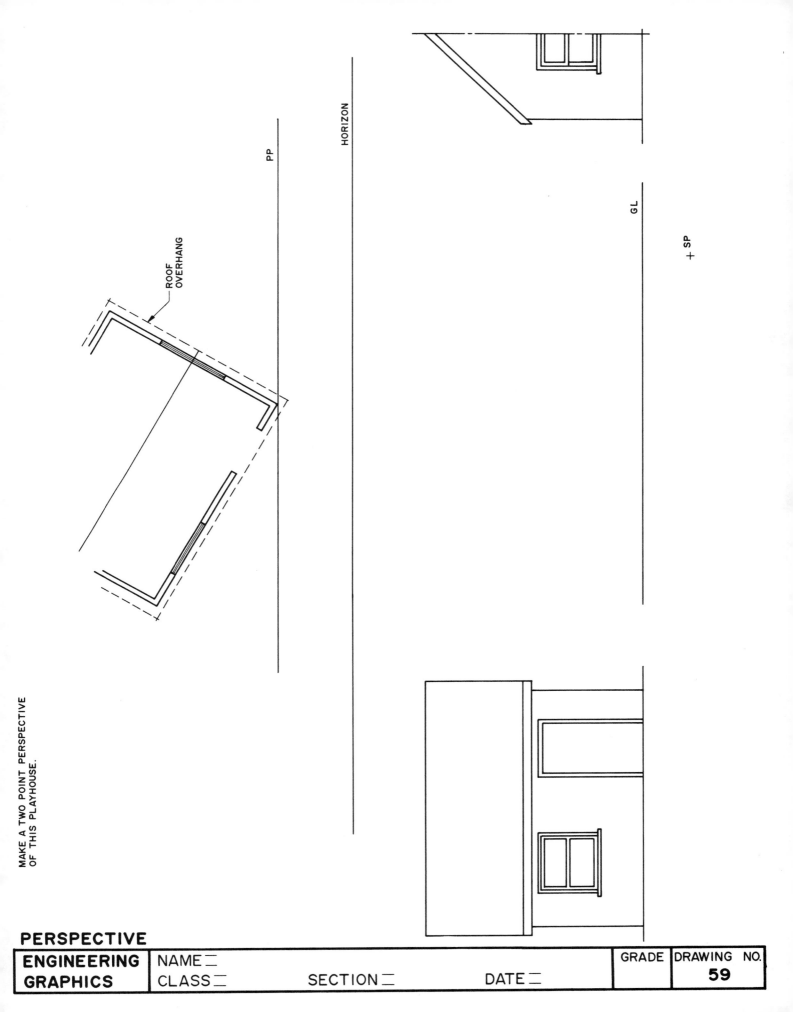

MAKE A TWO POINT PERSPECTIVE
OF THIS PLAYHOUSE.

ROOF
OVERHANG

PP

HORIZON

GL

+ SP

**PERSPECTIVE**

| ENGINEERING | NAME = | | | GRADE | DRAWING NO. |
|---|---|---|---|---|---|
| GRAPHICS | CLASS = | SECTION = | DATE = | | **59** |

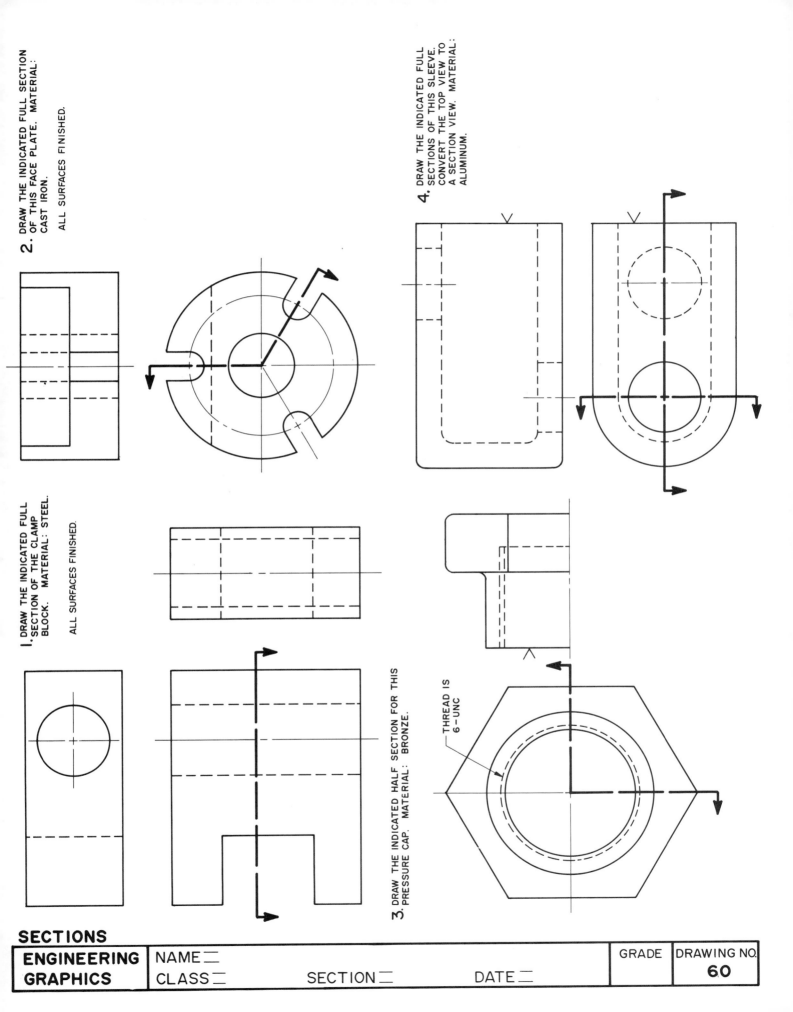

2. DRAW THE INDICATED FULL SECTION OF THIS FACE PLATE. MATERIAL: CAST IRON.

ALL SURFACES FINISHED.

4. DRAW THE INDICATED FULL SECTIONS OF THIS SLEEVE. CONVERT THE TOP VIEW TO A SECTION VIEW. MATERIAL: ALUMINUM.

1. DRAW THE INDICATED FULL SECTION OF THE CLAMP BLOCK. MATERIAL: STEEL.

ALL SURFACES FINISHED.

3. DRAW THE INDICATED HALF SECTION FOR THIS PRESSURE CAP. MATERIAL: BRONZE.

THREAD IS 6-UNC

**SECTIONS**

| **ENGINEERING** | NAME ⎓ | | | GRADE | DRAWING NO. |
| **GRAPHICS** | CLASS ⎓ | SECTION ⎓ | DATE ⎓ | | **60** |

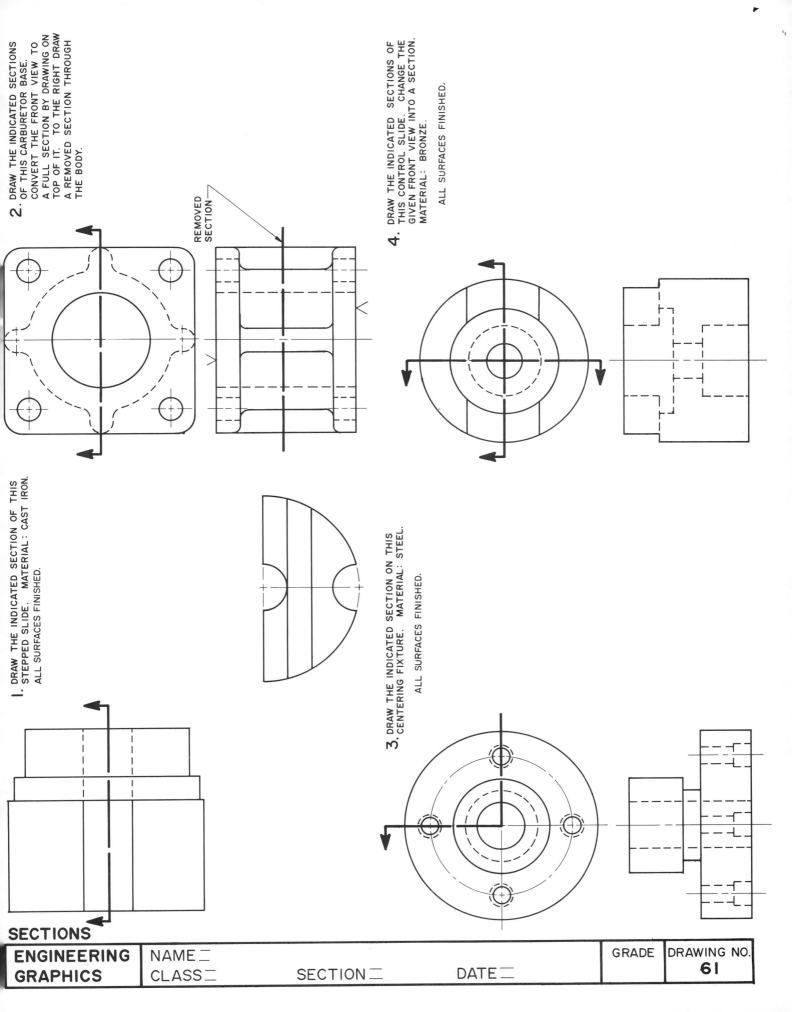

2. DRAW THE INDICATED SECTIONS
OF THIS CARBURETOR BASE.
CONVERT THE FRONT VIEW TO
A FULL SECTION BY DRAWING ON
TOP OF IT. TO THE RIGHT DRAW
A REMOVED SECTION THROUGH
THE BODY.

REMOVED
SECTION

4. DRAW THE INDICATED SECTIONS OF
THIS CONTROL SLIDE. CHANGE THE
GIVEN FRONT VIEW INTO A SECTION.
MATERIAL: BRONZE.

ALL SURFACES FINISHED.

1. DRAW THE INDICATED SECTION OF THIS
STEPPED SLIDE. MATERIAL: CAST IRON.
ALL SURFACES FINISHED.

3. DRAW THE INDICATED SECTION ON THIS
CENTERING FIXTURE. MATERIAL: STEEL.
ALL SURFACES FINISHED.

**SECTIONS**

| ENGINEERING | NAME | | | GRADE | DRAWING NO. |
|---|---|---|---|---|---|
| GRAPHICS | CLASS | SECTION | DATE | | 61 |

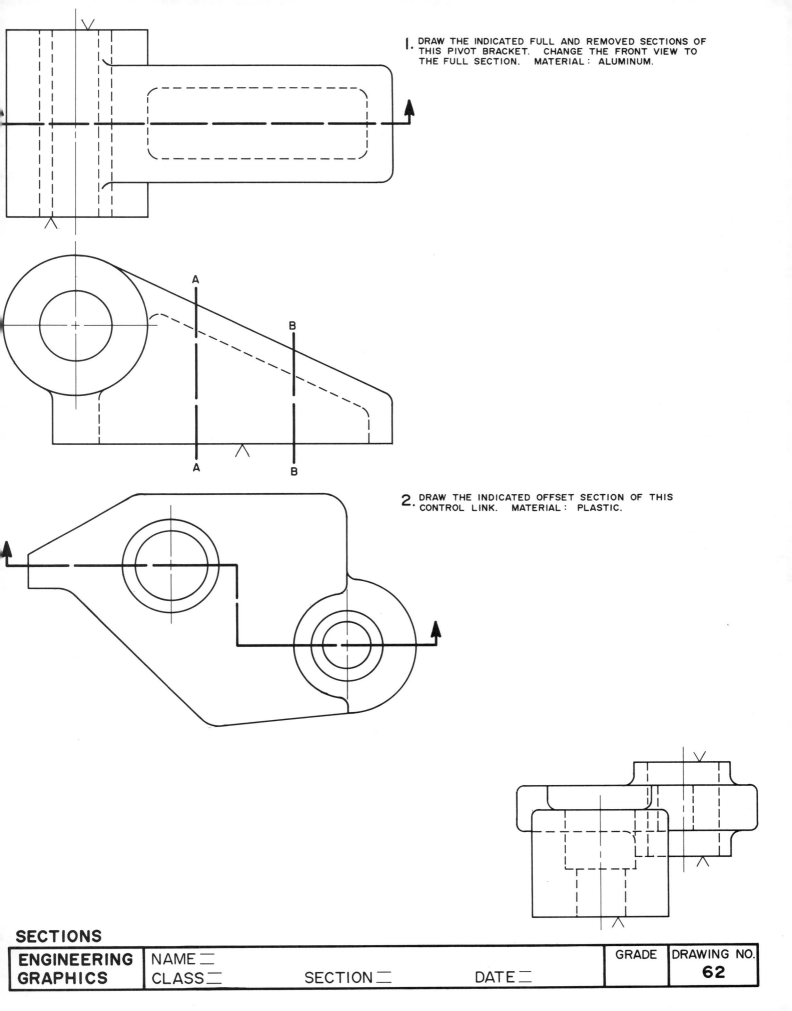

1. DRAW THE INDICATED FULL AND REMOVED SECTIONS OF THIS PIVOT BRACKET. CHANGE THE FRONT VIEW TO THE FULL SECTION. MATERIAL: ALUMINUM.

2. DRAW THE INDICATED OFFSET SECTION OF THIS CONTROL LINK. MATERIAL: PLASTIC.

A

B

A

B

**SECTIONS**

| **ENGINEERING** **GRAPHICS** | NAME ▭ CLASS ▭ | SECTION ▭ | DATE ▭ | GRADE | DRAWING NO. **62** |

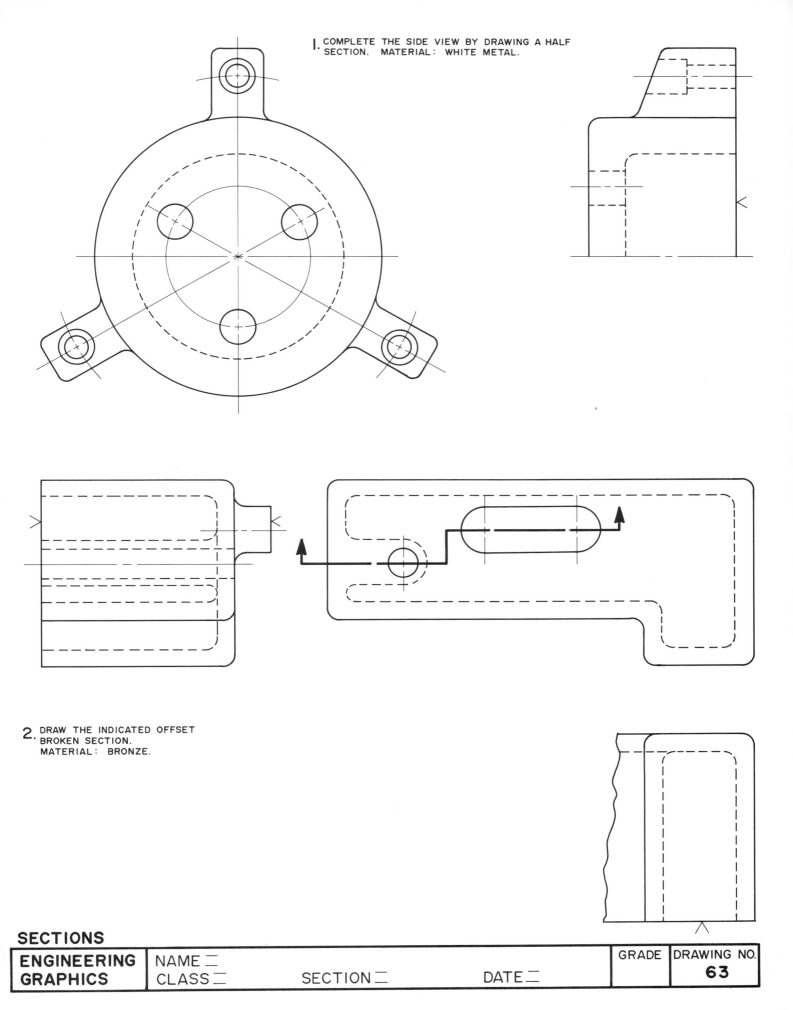

1. COMPLETE THE SIDE VIEW BY DRAWING A HALF
SECTION. MATERIAL: WHITE METAL.

2. DRAW THE INDICATED OFFSET
BROKEN SECTION.
MATERIAL: BRONZE.

**SECTIONS**

| ENGINEERING GRAPHICS | NAME ― CLASS ― | SECTION ― | DATE ― | GRADE | DRAWING NO. 63 |
|---|---|---|---|---|---|

**1.** DRAW THE INDICATED REVOLVED SECTIONS.  MATERIAL: STEEL.

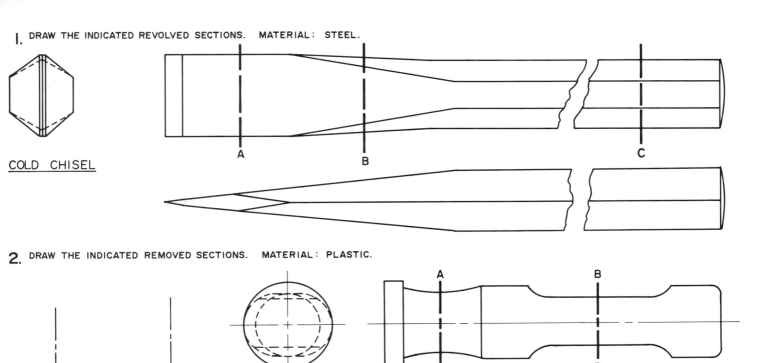

COLD CHISEL

A        B        C

**2.** DRAW THE INDICATED REMOVED SECTIONS.  MATERIAL: PLASTIC.

A-A        B-B

A        B

A        B

HANDLE FOR A WOOD CHISEL

**3.** DRAW THE INDICATED AUXILIARY
SECTION.  MATERIAL: ALUMINUM.

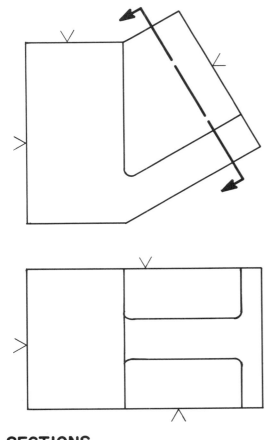

**4.** DRAW A BROKEN—OUT SECTION ON THE FRONT VIEW
SHOWING DETAILS OF THE INTERIOR HUB.
MATERIAL: CAST IRON.

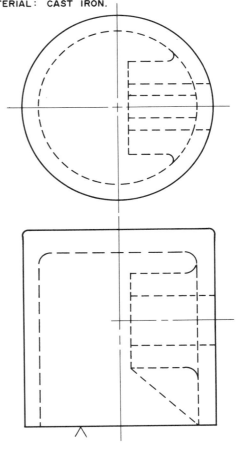

**SECTIONS**

| ENGINEERING | NAME ⊏ | | | GRADE | DRAWING NO. |
|---|---|---|---|---|---|
| **GRAPHICS** | CLASS ⊏ | SECTION ⊏ | DATE ⊏ | | **64** |

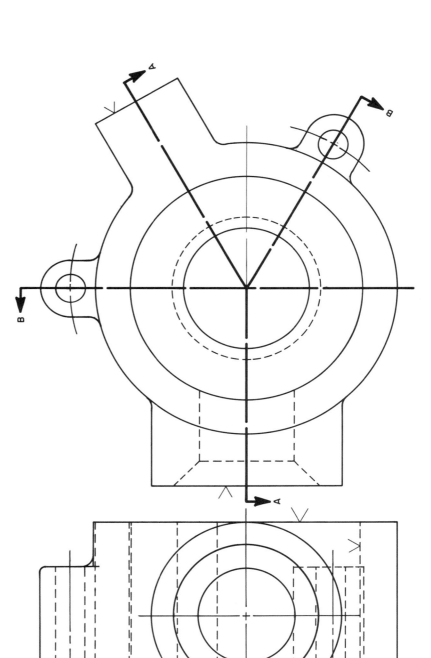

SECTION A – A

DRAW THE INDICATED SECTIONS.

**SECTIONS**

| ENGINEERING | NAME⌐ | | | GRADE | DRAWING NO. |
|---|---|---|---|---|---|
| GRAPHICS | CLASS⌐ | SECTION⌐ | DATE⌐ | | **65** |

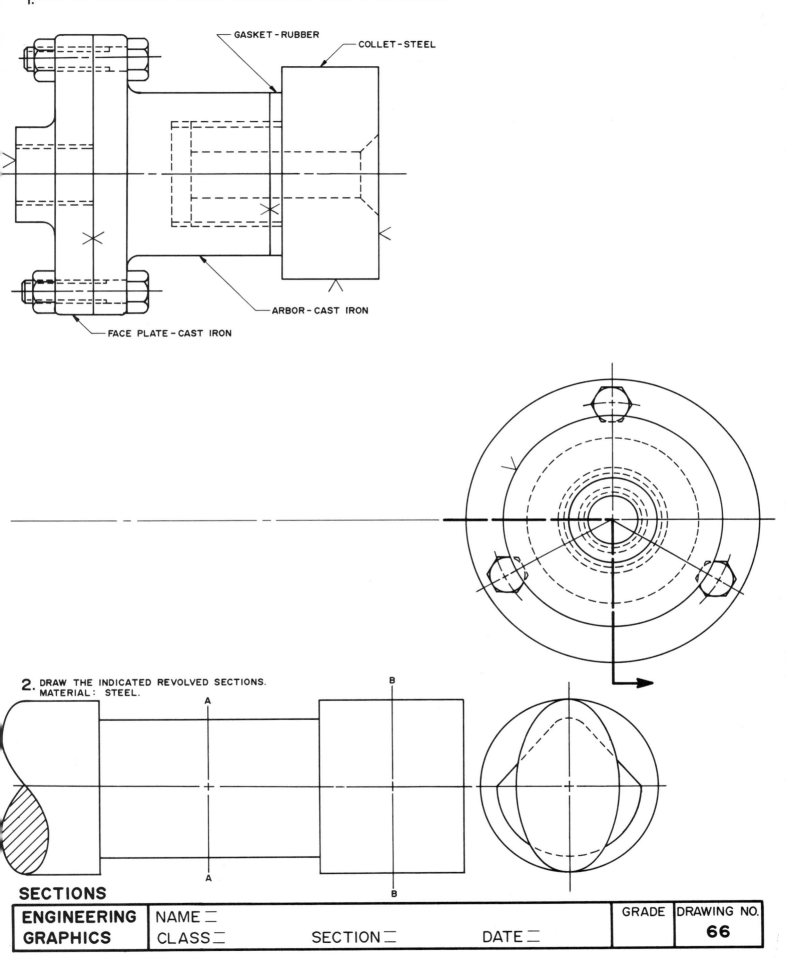

1. DRAW THE INDICATED HALF SECTION. MATERIALS ARE NOTED ON THE DRAWING.

GASKET - RUBBER

COLLET - STEEL

ARBOR - CAST IRON

FACE PLATE - CAST IRON

2. DRAW THE INDICATED REVOLVED SECTIONS.
MATERIAL: STEEL.

A

B

A

B

**SECTIONS**

| ENGINEERING | NAME ⎓ | | | GRADE | DRAWING NO. |
|---|---|---|---|---|---|
| GRAPHICS | CLASS ⎓ | SECTION ⎓ | DATE ⎓ | | **66** |

1. DIMENSION THIS PLATE USING THE ALIGNED
   SYSTEM. FIND THE SIZES BY MEASURING
   THE DRAWING IN INCHES. USE COMMON
   FRACTIONS. SCALE: FULL SIZE.

NOTE: ALL SURFACES
FINISHED.

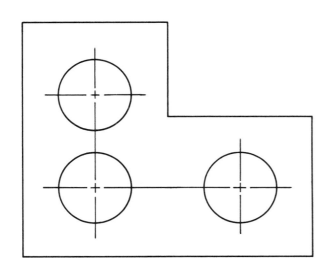

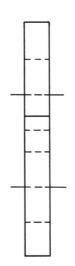

2. DIMENSION THIS CLAMP USING THE UNIDIRECTIONAL SYSTEM.
   FIND THE SIZES BY MEASURING THE DRAWING IN DECIMAL
   INCHES. SCALE: FULL SIZE.

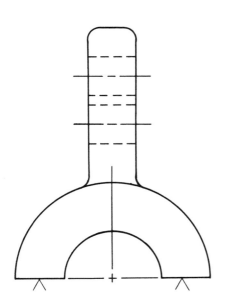

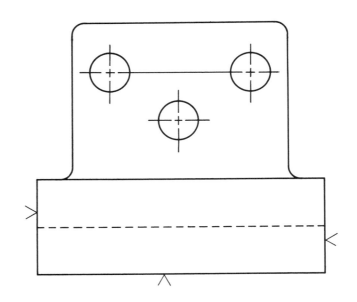

**DIMENSIONING**

| ENGINEERING GRAPHICS | NAME ⌐ CLASS ⌐ | SECTION ⌐ | DATE ⌐ | GRADE | DRAWING NO. 67 |

**I.** DIMENSION IN MILLIMETERS. MEASURE THE DRAWING TO FIND THE SIZES. SCALE: FULL SIZE.

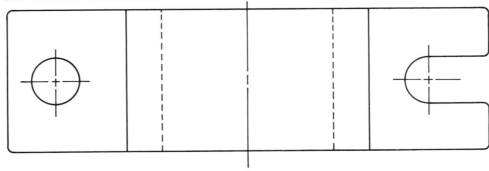

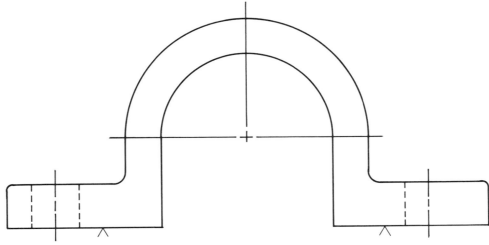

**2.** DIMENSION IN DECIMAL INCHES. MEASURE THE DRAWING TO FIND THE SIZE. SCALE: FULL SIZE.

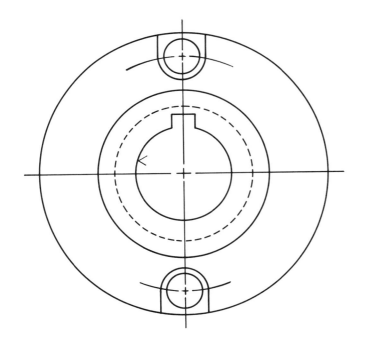

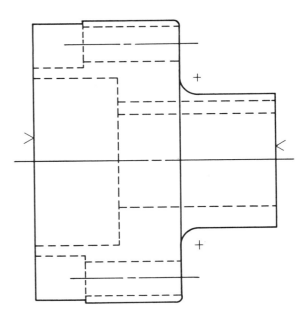

# DIMENSIONING

| ENGINEERING GRAPHICS | NAME ⹀ CLASS ⹀ | SECTION ⹀ | DATE ⹀ | GRADE | DRAWING NO. 68 |

1. DIMENSION IN DECIMAL INCHES. USE THE UNIDIRECTIONAL SYSTEM. MEASURE THE DRAWING TO FIND SIZES. SCALE: $\frac{1}{2}$" = 1".

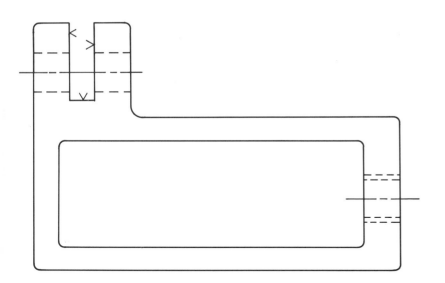

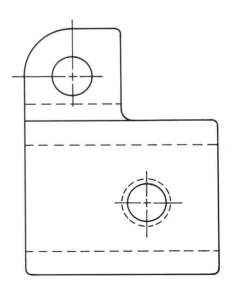

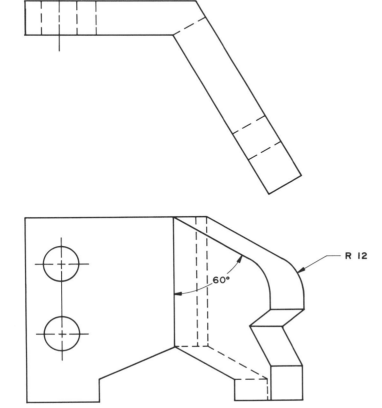

R 12

60°

2. DIMENSION IN MILLIMETERS. MEASURE THE DRAWING TO FIND THE SIZE. SCALE: 1:2. HINT: DRAW AN AUXILIARY VIEW OF THE INCLINED SURFACE.

ALL SURFACES FINISHED.

METRIC

**DIMENSIONING**

| ENGINEERING | NAME = | | | GRADE | DRAWING NO. |
|---|---|---|---|---|---|
| GRAPHICS | CLASS = | SECTION = | DATE = | | **69** |

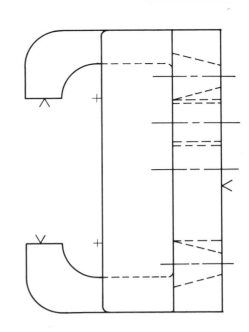

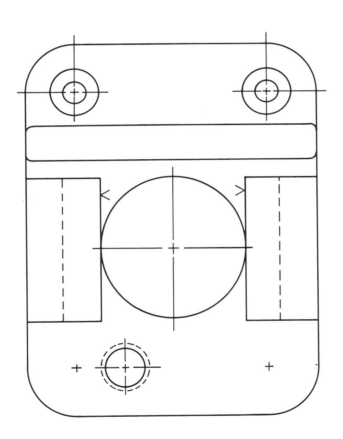

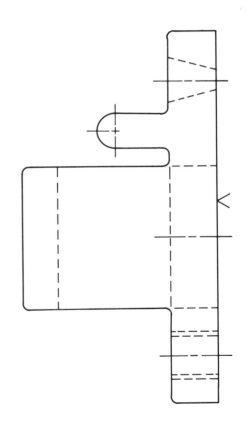

# DIMENSIONING

| ENGINEERING | NAME ⎓ | | | GRADE | DRAWING NO. |
|---|---|---|---|---|---|
| GRAPHICS | CLASS ⎓ | SESTION ⎓ | DATE ⎓ | | **70** |

1. DIMENSION IN MILLIMETERS. COUNTERSINK IS 82°. USE EIGHT POINTS TO DIMENSION THE IRREGULAR CURVE. MEASURE THE DRAWING TO FIND THE SIZE. SCALE: 1:1.

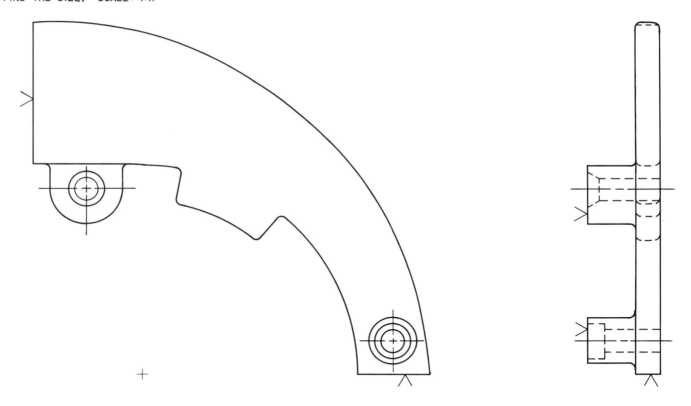

2. DIMENSION IN DECIMAL INCHES. MEASURE THE DRAWING TO FIND SIZES. SCALE: FULL SIZE.

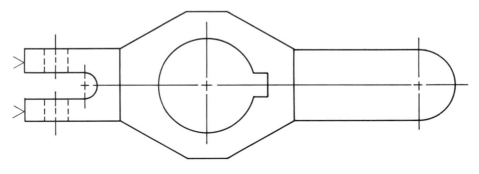

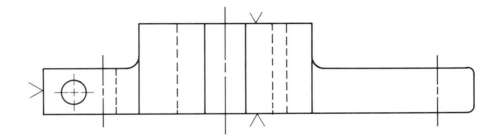

**DIMENSIONING**

| ENGINEERING GRAPHICS | NAME ▭ | | | GRADE | DRAWING NO. |
|---|---|---|---|---|---|
| | CLASS ▭ | SECTION ▭ | DATE ▭ | | 71 |

APPLY THE INDICATED AMERICAN STANDARD LIMITS AND FITS TO
THE HOLES AND SHAFTS BELOW. RECORD THE DATA IN THOUSANDTHS
IN THE TABLE BELOW AND DIMENSION THE SHAFTS AND HOLES.

USING THE BASIC SHAFT SYSTEM APPLY AM. NAT. STD.
TOLERANCES TO THE HOLE AND SHAFT.  SHAFT: BASIC
SIZE 1.500.

RC 6

LC 4

FN 4

| CLASS OF FIT | RC 6-1.000 | LT 2-1.500 | FN 4-.750 |
|---|---|---|---|
| LIMITS OF CLEARANCE OR INTERFERANCE | | | |
| HOLE TOLERANCE | | | |
| SHAFT TOLERANCE | | | |
| ALLOWANCE AT MMC | | | |

LOCATE THE CENTER OF EACH HOLE FROM DATUM A WITH
LIMITS OF +.010 AND +.020.   MEASURE THE DRAWING
TO FIND THE DISTANCES.

DATUM A

DIMENSION THE ANGLE BELOW USING THE BASIC ANGLE METHOD.
THE TOLERANCES FOR THE LOCATION OF THE ENDS OF THE ANGLE
ARE +.010 INCHES.   MEASURE THE DRAWING TO FIND THE SIZES.

DIMENSION THE TAPER BELOW BY GIVING THE BASIC TAPER
AND THE BASIC DIAMETER.   THE TAPER ON THE DIAMETER
IS +.125 PER INCH.   MEASURE THE DRAWING TO FIND THE
BASIC DIAMETER.

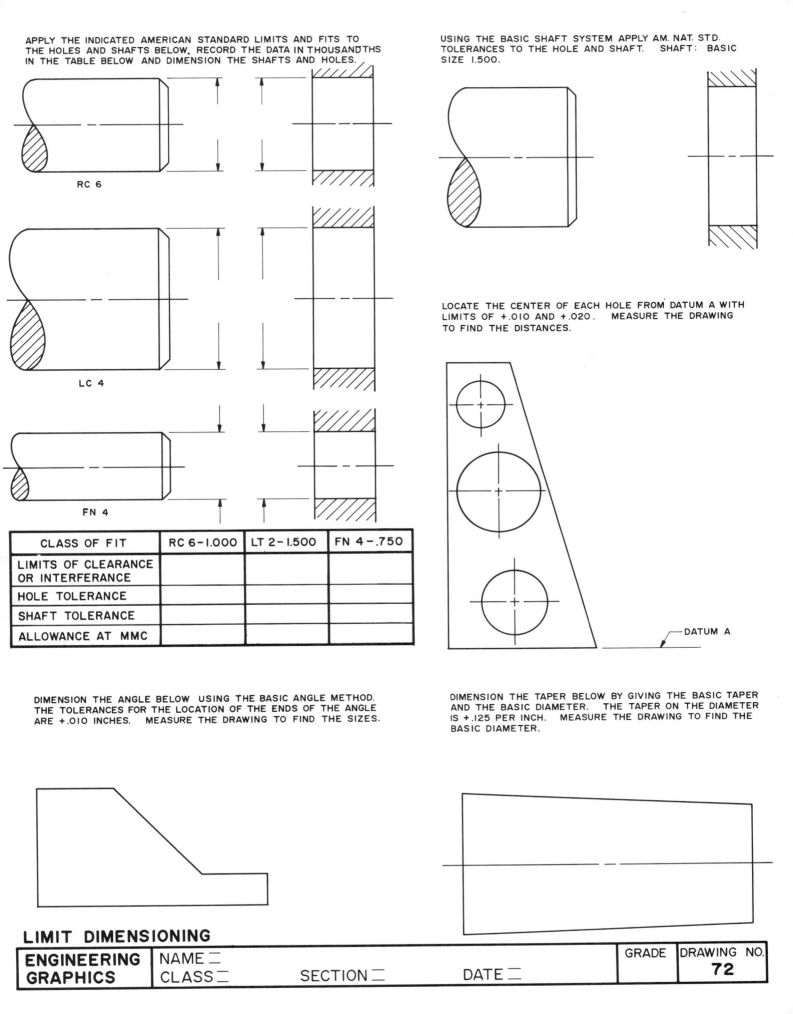

## LIMIT DIMENSIONING

| ENGINEERING GRAPHICS | NAME = CLASS = | SECTION = | DATE = | GRADE | DRAWING NO. 72 |
|---|---|---|---|---|---|

1. USING THE ISO SYSTEM OF LIMITS AND FITS DIMENSION THE HOLES AND SHAFTS BELOW. USE THE SIZES ON EACH DRAWING AND RECORD THE DATA IN THE TABLE BELOW.

2. USING THE PREFERRED SHAFT ISO SYSTEM OF LIMITS AND FITS DIMENSION THE SLOT AND GUIDE PIN BELOW. THE BASIC SIZE ON THE WIDTH OF THE GUIDE IS 40 mm AND THE FIT IS G7/h6. THE BASIC SIZE OF THE DEPTH OF PIN IS 16 mm AND THE FIT IS C11/h11.

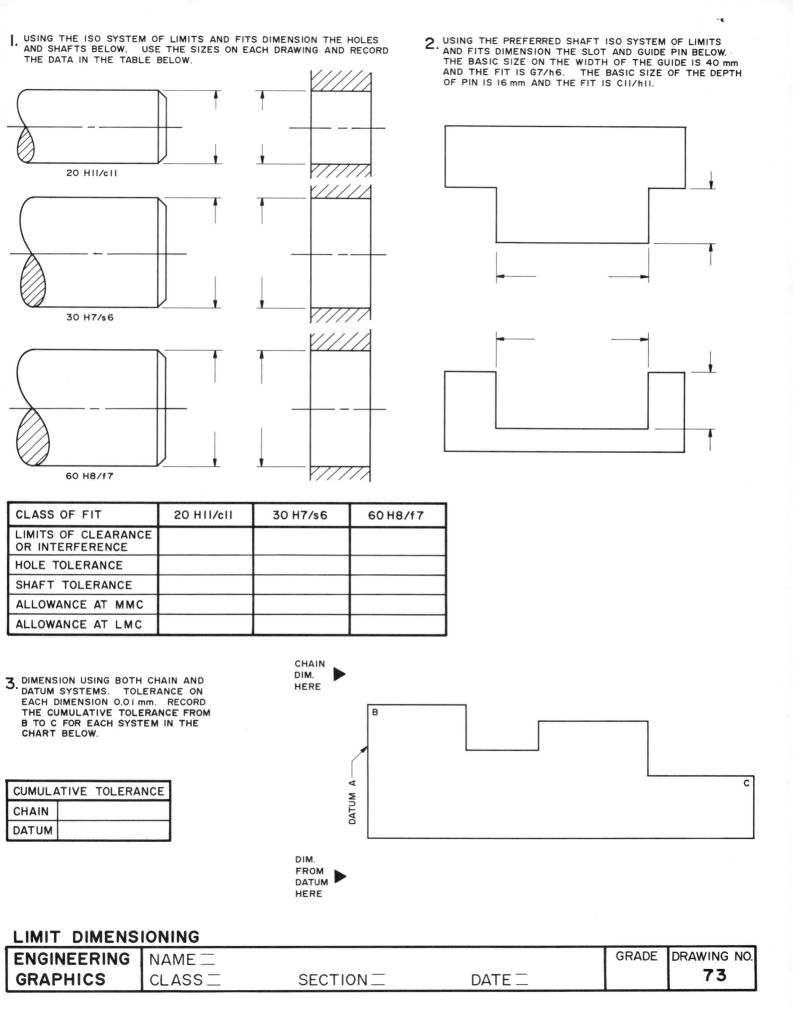

20 H11/c11

30 H7/s6

60 H8/f7

| CLASS OF FIT | 20 H11/c11 | 30 H7/s6 | 60 H8/f7 |
|---|---|---|---|
| LIMITS OF CLEARANCE OR INTERFERENCE | | | |
| HOLE TOLERANCE | | | |
| SHAFT TOLERANCE | | | |
| ALLOWANCE AT MMC | | | |
| ALLOWANCE AT LMC | | | |

3. DIMENSION USING BOTH CHAIN AND DATUM SYSTEMS. TOLERANCE ON EACH DIMENSION 0.01 mm. RECORD THE CUMULATIVE TOLERANCE FROM B TO C FOR EACH SYSTEM IN THE CHART BELOW.

CHAIN DIM. HERE ▶

| CUMULATIVE TOLERANCE | |
|---|---|
| CHAIN | |
| DATUM | |

B

DATUM A

C

DIM. FROM DATUM HERE ▶

## LIMIT DIMENSIONING

| ENGINEERING GRAPHICS | NAME ▭  CLASS ▭  SECTION ▭  DATE ▭ | GRADE | DRAWING NO. 73 |
|---|---|---|---|

# TOLERANCES OF LOCATION

**1.** USE A FEATURE CONTROL SYMBOL TO SHOW THE TRUE POSITION OF THIS HOLE WITHIN A TOLERANCE ZONE OF .010" DIAMETER.

**2.** USE A FEATURE CONTROL SYMBOL TO SPECIFY THE CONCENTRICITY OF CYLINDERS A AND B WITHIN A TOLERANCE ZONE OF .025. USE CYLINDER A AS THE DATUM.

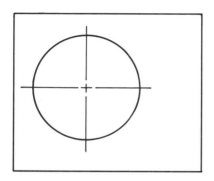

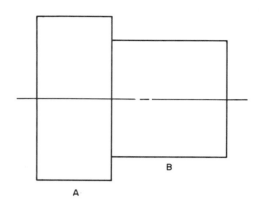

**3.** USE A FEATURE CONTROL SYMBOL TO SPECIFY SYMMETRY OF THE SLOT REGARDLESS OF FEATURE SIZE WHEN THE TOLERANCE ZONE IS 0.05 mm. THE DATUM SURFACE IS A.

**4.** USE A FEATURE CONTROL SYMBOL TO SHOW THE TRUE POSITION OF THE HOLES WITHIN A TOLERANCE ZONE OF 0.01 mm.

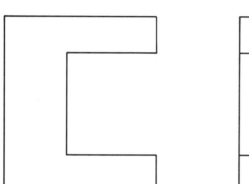 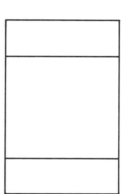

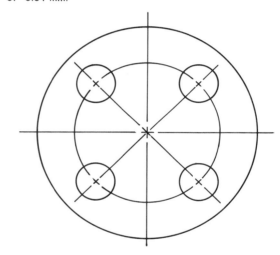

# TOLERANCES OF FORM

**1.** USE A FEATURE CONTROL SYMBOL TO SPECIFY STRAIGHTNESS FOR CYLINDER B WITHIN A TOLERANCE ZONE OF .025".

**2.** USE A FEATURE CONTROL SYMBOL TO SPECIFY FLATNESS FOR SURFACE A AND B. SURFACE A HAS A TOLERANCE ZONE OF .025" AND B .050."

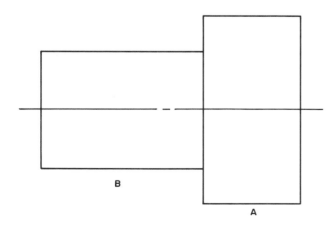

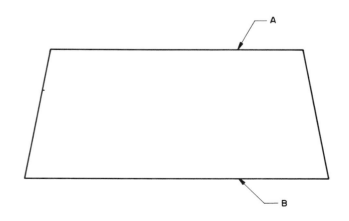

# GEOMETRIC TOLERANCING

| ENGINEERING | NAME ⊐ | | | | GRADE | DRAWING NO. |
| GRAPHICS | CLASS ⊐ | SECTION ⊐ | DATE ⊐ | | | **74** |

# TOLERANCES OF FORM

1. USE A FEATURE CONTROL SYMBOL TO SPECIFY ROUNDNESS OF THE CYLINDER BELOW WITHIN A TOLERANCE ZONE OF .025".

2. USE A FEATURE CONTROL SYMBOL TO SPECIFY CYLINDRICITY OF CYLINDERS A AND B. CYLINDER A HAS A TOLERANCE ZONE OF 0.01 mm AND B 0.02 mm.

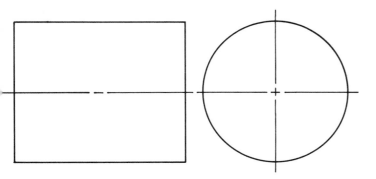
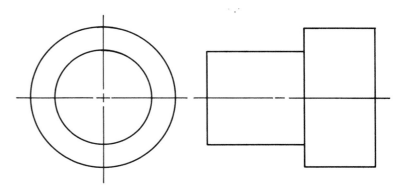

3. USE A FEATURE CONTROL SYMBOL TO SPECIFY BILATERAL PROFILE TOLERANCE OF THE CURVED SURFACE BELOW WITHIN A TOLERANCE ZONE OF .030". THE DATUM PLANE IS A.

4. USE A FEATURE CONTROL SYMBOL TO SPECIFY ANGULARITY FOR SURFACE A BELOW WITHIN A TOLERANCE ZONE OF .050". SURFACE B IS THE DATUM PLANE.

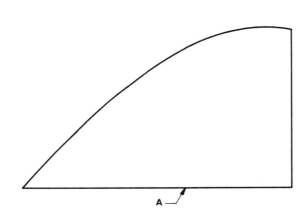
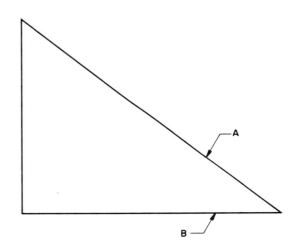

5. USE A FEATURE CONTROL SYMBOL TO SPECIFY PARALLELISM FOR SURFACES A AND B WITHIN A TOLERANCE ZONE OF 0.50 mm. SURFACE C IS THE DATUM PLANE.

6. USE A FEATURE CONTROL SYMBOL TO SPECIFY PERPENDICULARITY FOR SURFACES A AND B WITHIN A TOLERANCE ZONE OF .040". SURFACE C IS THE DATUM PLANE.

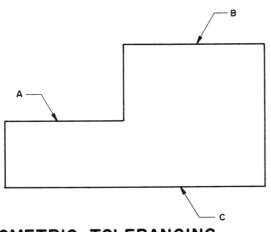
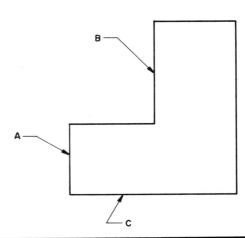

# GEOMETRIC TOLERANCING

| ENGINEERING | NAME = | | | GRADE | DRAWING NO. |
| GRAPHICS | CLASS = | SECTION = | DATE = | | 75 |

DRAW THE THREADS INDICATED BELOW AND DIMENSION THEM WITH A PROPER THREAD NOTE. USE THE AMERICAN NATIONAL THREAD TABLES TO SECURE NEEDED DATA.

1. EXTERNAL THREAD, DIAMETER $\frac{5}{8}$", UNC, THREAD LENGTH 2", CLASS 2, SIMPLIFIED REPRESENTATION.

2. EXTERNAL THREAD, DIAMETER $\frac{3}{4}$", UNF, THREAD LENGTH 2", CLASS 2, SCHEMATIC REPRESENTATION.

END VIEW            SIDE VIEW

3. DRAW THE END AND SIDE VIEWS OF THE FOLLOWING INTERNAL THREAD: DIAMETER .75", UNC, CLASS 2, THROUGH THREAD, SIMPLIFIED REPRESENTATION.

END VIEW            SIDE

4. DRAW THE END AND SIDE VIEWS OF THE FOLLOWING THREAD: DIAMETER 1", UNF, CLASS 3, THRU THREAD, DETAILED REPRESENTATION.

5. DRAW THE SIDE VIEW OF AN EXTERNAL THREAD, DIAMETER 1.5, UNC CLASS 2, 3" LONG, DETAILED REPRESENTATION.

SIMPLIFIED

SCHEMATIC

DETAILED

6. DRAW THE FOLLOWING THREAD IN SECTION ABOVE: DRILLED HOLE DIA. .040", 2.5" DEEP, .5 DIA., UNC THREAD, 2.0 DEEP.

7. DRAW AN EXTERNAL ACME THREAD, DIAMETER 2" THREAD 3" LONG, IN THE SPACE AT THE LEFT.

# THREADS AND FASTENERS

| ENGINEERING GRAPHICS | NAME ⹀<br>CLASS ⹀ | SECTION ⹀ | DATE ⹀ | GRADE | DRAWING NO.<br>76 |

**1.** WRITE THE FOLLOWING METRIC THREAD NOTES:

| DATA | THREAD NOTE |
|---|---|
| **A.** EXTERNAL GENERAL PURPOSE THREAD, DIAMETER 8 mm, PITCH 1. | **A.** ___ |
| **B.** INTERNAL THREAD, PRECISION, DIAMETER 6 mm AFTER PLATING, PITCH 1. | **B.** ___ |
| **C.** THREADED HOLE, DRILL 40 mm DEEP, THREAD 35 mm DEEP, GENERAL PURPOSE THREAD, DIAMETER 10 mm, PITCH 1.5 mm. | **C.** ___ |

**2.** FIND THE PITCHES FOR THE FOLLOWING METRIC THREADS. RECORD BELOW.

| DIAMETER | COARSE | FINE |
|---|---|---|
| 4 | | |
| 8 | | |
| 12 | | |
| 16 | | |
| 20 | | |

**3.** FIND THE NUMBER OF THREADS PER INCH FOR THE FOLLOWING THREADS. RECORD BELOW.

| DATA | THREADS PER INCH |
|---|---|
| METRIC, DIAMETER 30 mm. | |
| BUTTRESS, DIA. 3" PREFERRED SIZES | |
| STUB ACME, DIAMETER $1\frac{3}{8}$". | |
| ACME, DIAMETER $\frac{3}{4}$". | |
| UNIFIED NATIONAL FINE, DIA. $\frac{7}{8}$". | |

**4.** WRITE THREAD NOTES FOR THE FOLLOWING:

| DATA | THREAD NOTE |
|---|---|
| ACME, DIA. 1.25", 5 THREADS PER INCH. | |
| METRIC, DIA. 12 mm, PITCH 1.25 mm, EXTERNAL. | |
| AMERICAN NATIONAL STANDARD, UNIFIED NATIONAL FINE, DIA. 1.125", 10 THREADS, PER INCH, CLASS 2, EXTERNAL. | |
| BUTTRESS, DIA. 1.25", 10 THREADS, CLASS 2 FIT, EXTERNAL. | |

**5.** DRAW THE TOP AND SECTION VIEW OF THESE FASTENERS. USE SCHEMATIC THREAD REPRESENTATION.

$\frac{1}{2}$" STUD WITH HEX NUT, STUD $1\frac{3}{4}$" LONG, THD. $\frac{3}{4}$" EACH END.

$\frac{1}{2}$" HEX HEAD CAP SCREW, 1" LONG.

$\frac{1}{2}$" DIA. HEX HEAD BOLT AND NUT, $1\frac{7}{8}$" LONG.

NO. 10 HEX SOCKET CAP SCREW, 1" LONG.

M 10 X 1.5 METRIC SLOTTED OVAL HEAD MACHINE SCREW 25 mm. LONG.

# THREADS AND FASTENERS

| ENGINEERING GRAPHICS | NAME | | | GRADE | DRAWING NO. |
|---|---|---|---|---|---|
| | CLASS | SECTION | DATE | | **77** |

1. DRAW A DETAILED REPRESENTATION OF THE FOLLOWING COMPRESSION SPRING: OUTSIDE DIAMETER 60 mm, PITCH 11mm, WIRE DIAMETER 6 mm, FREE LENGTH 175 mm, OPEN GROUND END.

2. DRAW A SIMPLIFIED REPRESENTATION OF THE TWO SPRINGS BELOW.

10 COIL EXPANSION SPRING, FREE LENGTH 94 mm,
10 ACTIVE COILS, OUTSIDE DIA. 38 mm, WIRE DIA. 3 mm,
FULL LOOP.

TORSION SPRING, 8 COILS, WIRE DIA. $\frac{3}{16}$",
FREE LENGTH $1\frac{1}{4}$", OUTSIDE DIA. $1\frac{1}{2}$",
STRAIGHT WIRE ENDS $1\frac{1}{4}$" FROM CENTER LINE.

3. MAKE A SCHEMATIC REPRESENTATION BELOW OF THE TWO SPRINGS DETAILED ABOVE IN PROBLEM 2.

**SPRINGS**

| ENGINEERING | NAME = | | | GRADE | DRAWING NO. |
|---|---|---|---|---|---|
| GRAPHICS | CLASS = | SECTION = | DATE = | | **78** |

1. ON SHEET 80 MAKE A PATTERN DETAIL DRAWING FOR THE CYLINDER REAMER BASE AT THE RIGHT. SPECIFICATIONS INCLUDE DRAFT ANGLE 7°, MATERIAL CAST IRON, TOLERANCE ± .05", MACHINING ALLOWANCE .07". DRAW TO THE SCALE ½" = 1".

2. ON SHEET 81 MAKE A MACHINING DRAWING FOR THE CYLINDER REAMER BASE AT THE RIGHT. TOLERANCE ± .01". SURFACE FINISH – LAY CIRCULAR, REMOVE .07" TO PRODUCE FINISHED SURFACE, ROUGHNESS 1.6. DRAW TO THE SCALE ½" = 1".

3. ON SHEET 82 MAKE A MULTIDETAIL DRAWING OF THE SLIDING BOLT BELOW. ALL PARTS ARE ALUMINUM. THE BOTTOM OF THE BASE IS MACHINED WITH A TOLERANCE OF ± .01". DRAW FULL SIZE.

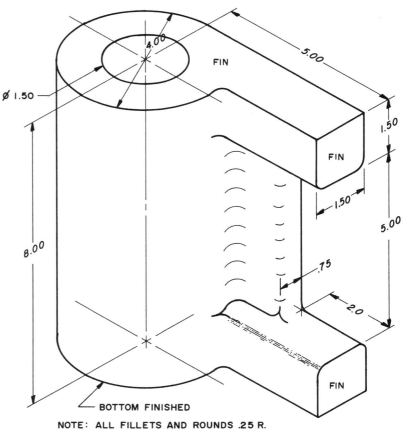

NOTE: ALL FILLETS AND ROUNDS .25 R.
ALL SURFACES TO BE FINISHED MARKED FIN.

## CYLINDER REAMER BASE

4. ON SHEET 83 MAKE A FORGING DRAWING OF THE HANDLE OF THE PISTON RING EXPANDER BELOW. SPECIFICATIONS INCLUDE DRAFT ANGLE 7°, WEIGHT OF FORGING 0.36 kg. SELECT SHRINKAGE, DIE-WEAR, MISMATCH, AND THICKNESS TOLERANCES, MACHINING ALLOWANCES, AND RADII FOR FILLETS AND ROUNDS FROM FORGING TABLES. SELECT A SCALE THAT WILL PERMIT THE DRAWING TO FIT ON THE SHEET.

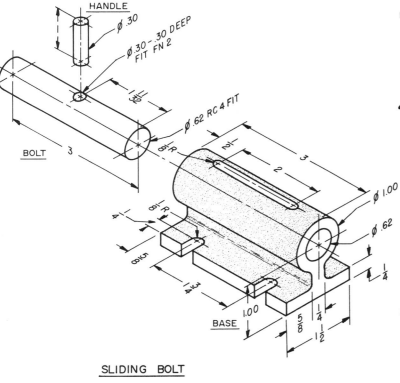

SLIDING BOLT

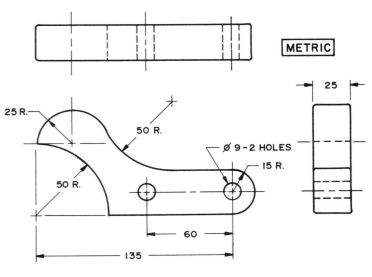

METRIC

PISTON RING EXPANDER HANDLE

# PRODUCTION DRAWINGS

| ENGINEERING | NAME = | | | GRADE | DRAWING NO. |
| GRAPHICS | CLASS = | SECTION = | DATE = | | 79 |

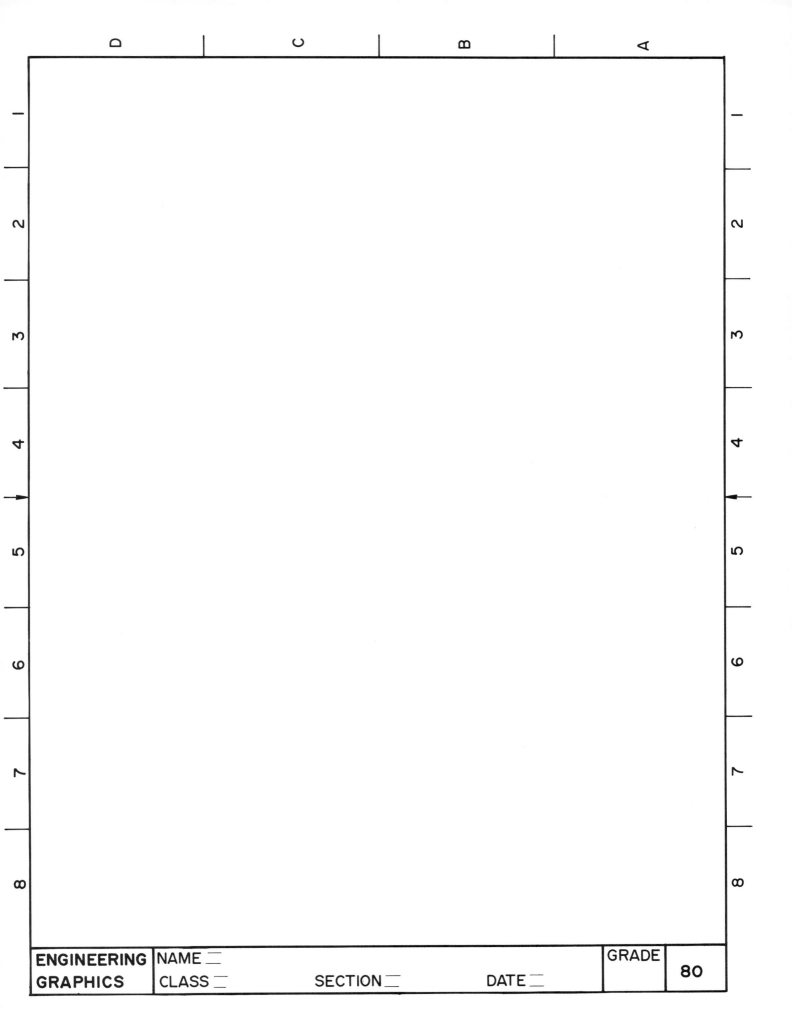

| D | C | B | A |
|---|---|---|---|

| ENGINEERING | NAME | | | GRADE | |
|---|---|---|---|---|---|
| GRAPHICS | CLASS | SECTION | DATE | | 80 |

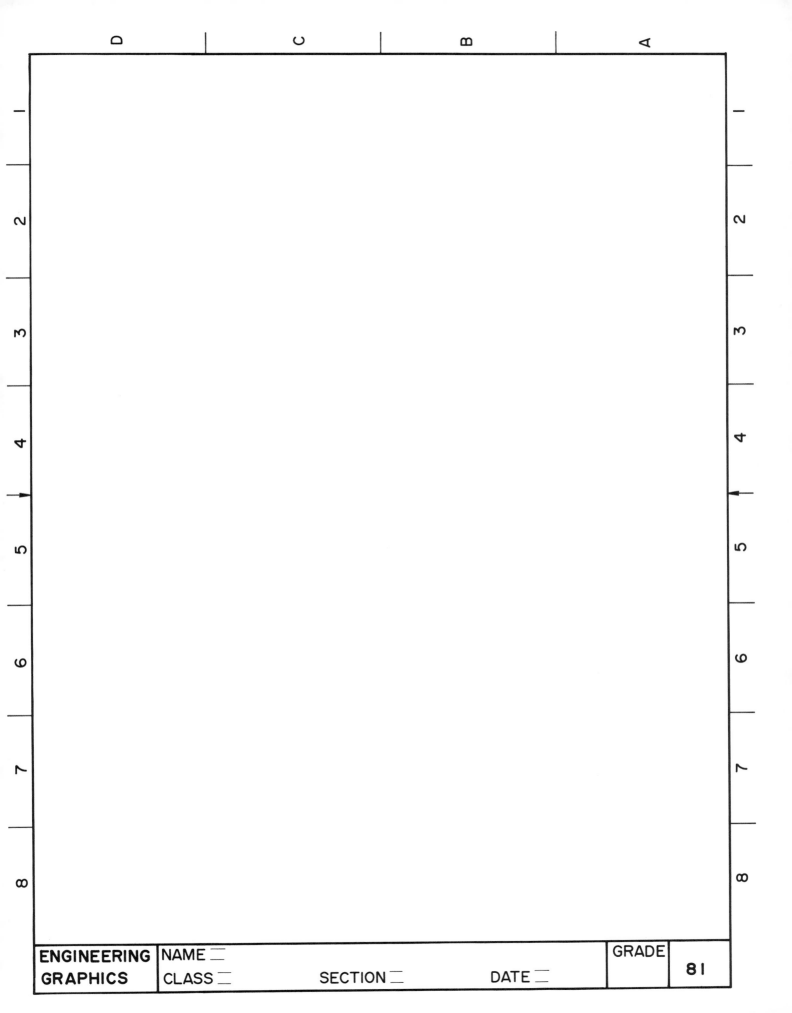

D      C      B      A

1    2    3    4    5    6    7    8

| ENGINEERING | NAME ▭ | | | GRADE | |
|---|---|---|---|---|---|
| GRAPHICS | CLASS ▭ | SECTION ▭ | DATE ▭ | | 81 |

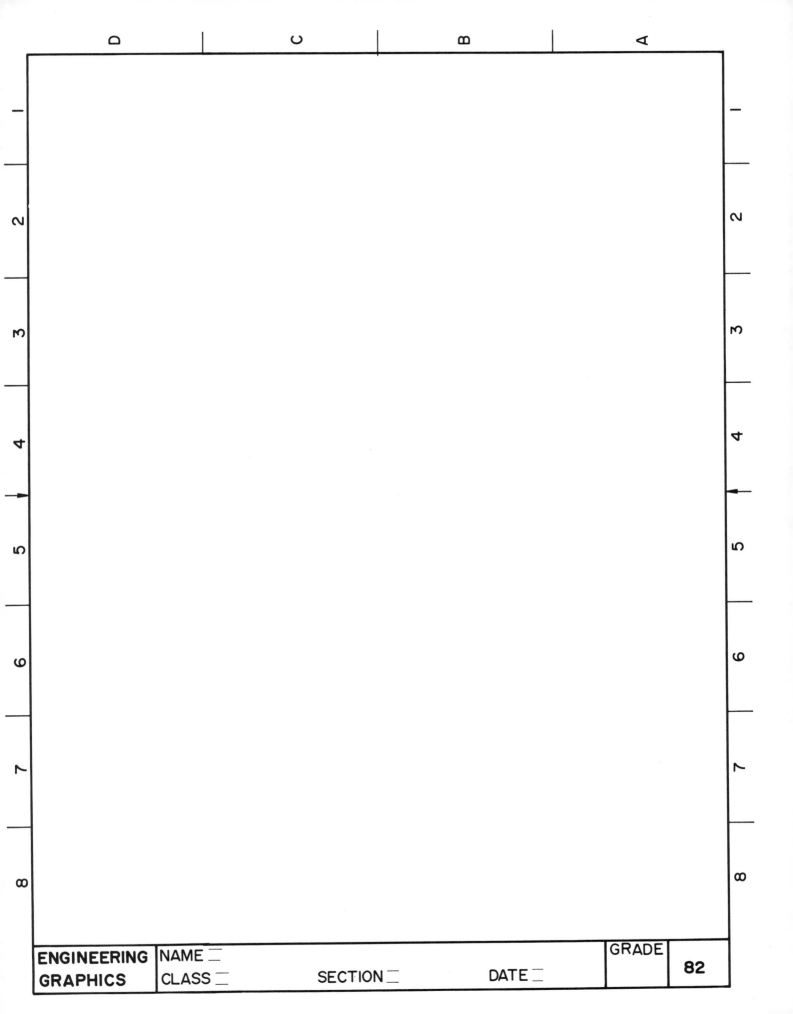

D     C     B     A

**ENGINEERING GRAPHICS**

NAME ⚊

CLASS ⚊     SECTION ⚊     DATE ⚊

GRADE

**82**

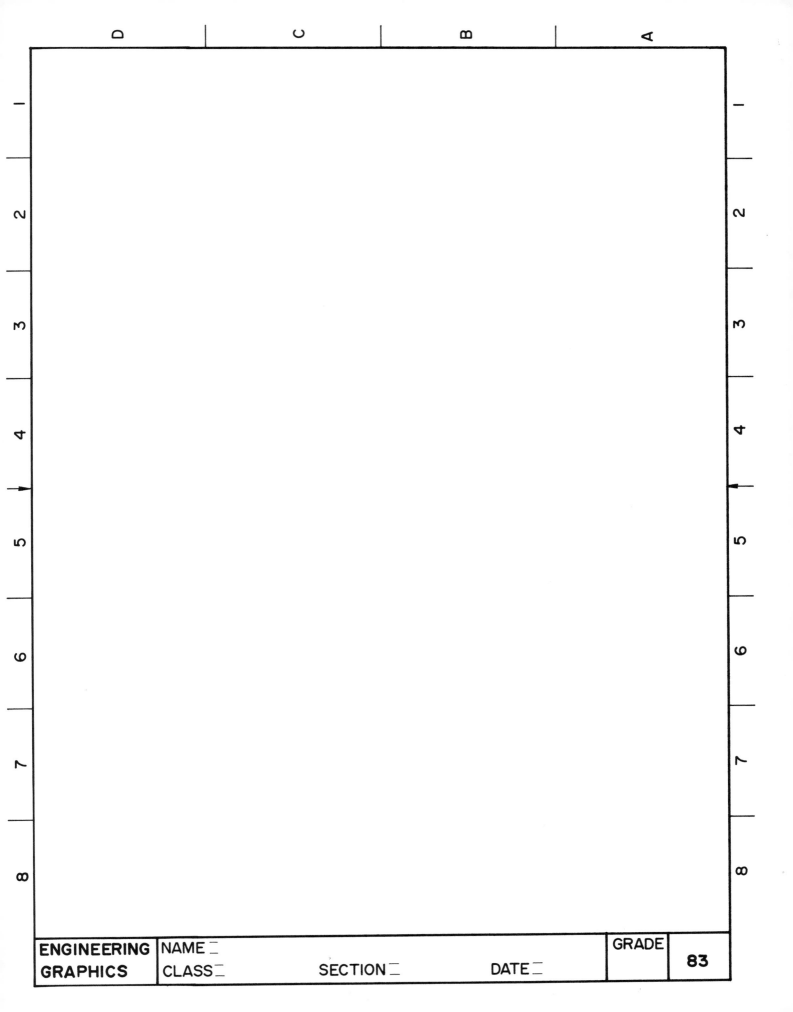

ENGINEERING
GRAPHICS

NAME
CLASS          SECTION          DATE

GRADE
83

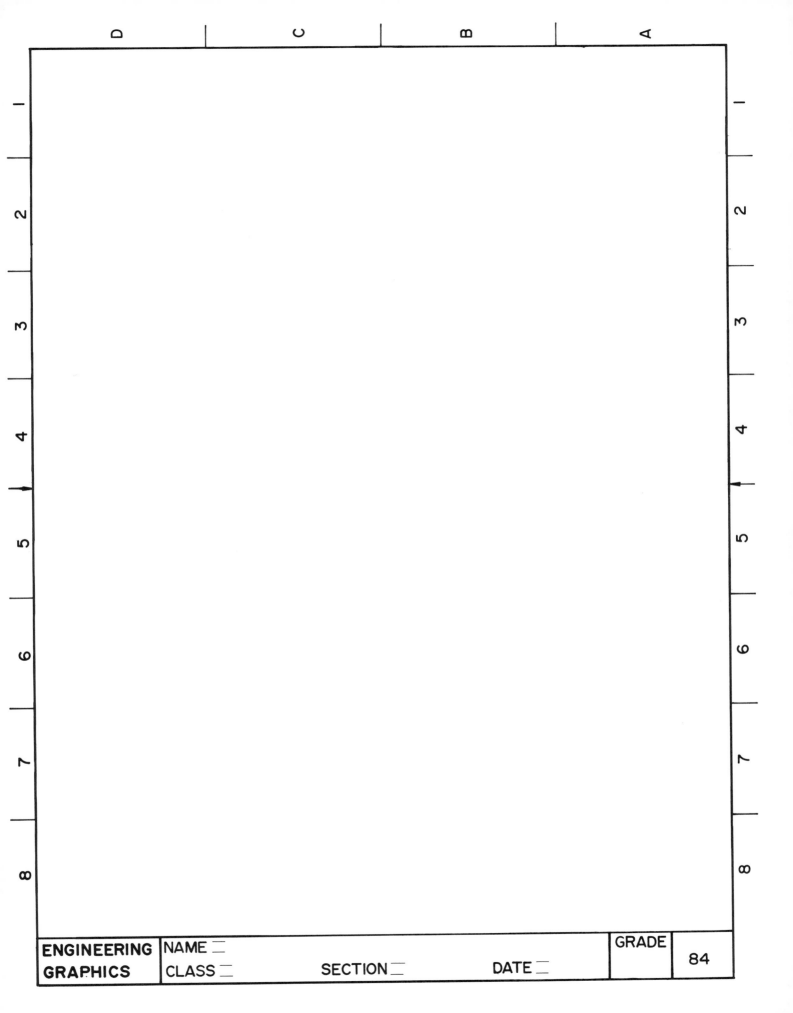

| D | C | B | A |
|---|---|---|---|

| ENGINEERING GRAPHICS | NAME ⎓<br>CLASS ⎓　　　SECTION ⎓　　　DATE ⎓ | GRADE | 84 |

ON SHEET 86 MAKE A STAMPING DETAIL DRAWING FOR EACH
PART OF THE FEED SEPARATION BLOCK AT THE RIGHT.
SCALE: ½" = 1".

ON SHEET 87 DEVELOP A PATTERN FOR EACH PART OF THE
FEED SEPARATION BLOCK.   SCALE: FULL SIZE.

ON SHEET 88 MAKE AN EXPLODED PICTORIAL ASSEMBLY OF
THE BACK-UP PLATE AND BASE SHOWN BELOW.   DRAW FULL
SIZE ON A SIZE B OR A3 SHEET.

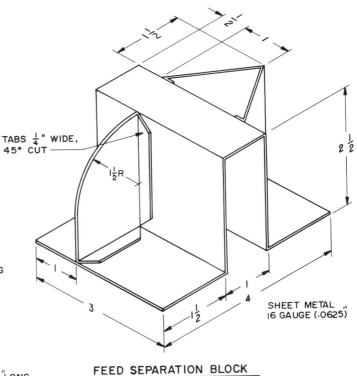

FEED SEPARATION BLOCK

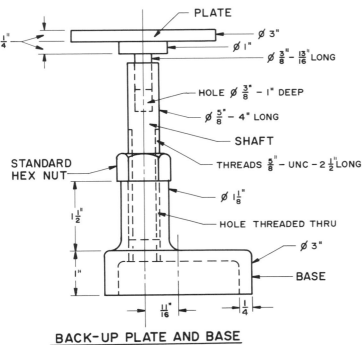

BACK-UP PLATE AND BASE

ON SHEET 89 MAKE DETAIL DRAWINGS OF EACH PART OF
THE HAND CHUCK AT THE RIGHT.   SCALE:  FULL SIZE.

ON SHEET 90 MAKE A GENERAL ASSEMBLY DRAWING OF
THE HAND CHUCK AT THE RIGHT.   SCALE:  FULL SIZE.

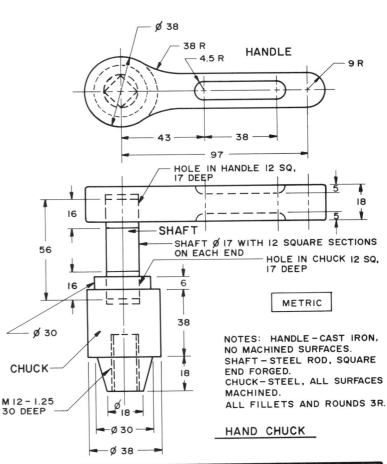

NOTES:  HANDLE – CAST IRON,
NO MACHINED SURFACES.
SHAFT – STEEL ROD, SQUARE
END FORGED.
CHUCK – STEEL, ALL SURFACES
MACHINED.
ALL FILLETS AND ROUNDS 3R.

HAND CHUCK

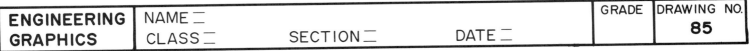

| ENGINEERING GRAPHICS | NAME ▭   CLASS ▭       SECTION ▭       DATE ▭ | GRADE | DRAWING NO. 85 |

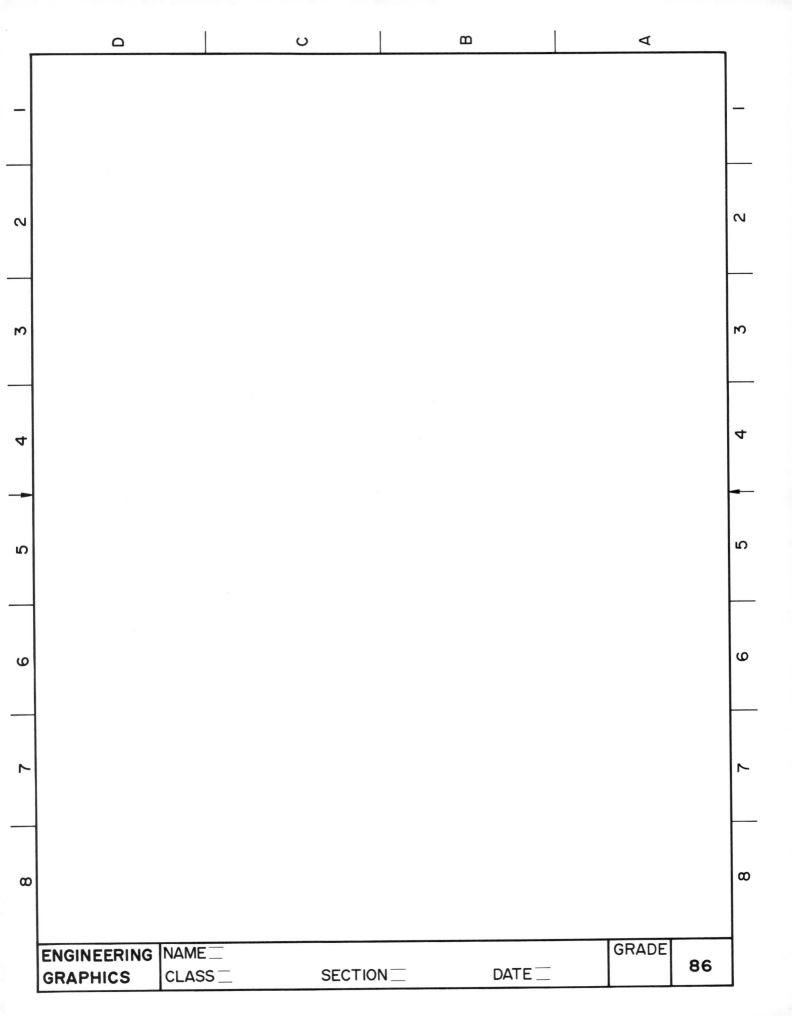

| ENGINEERING GRAPHICS | NAME | | GRADE | 86 |
| CLASS | SECTION | DATE | | |

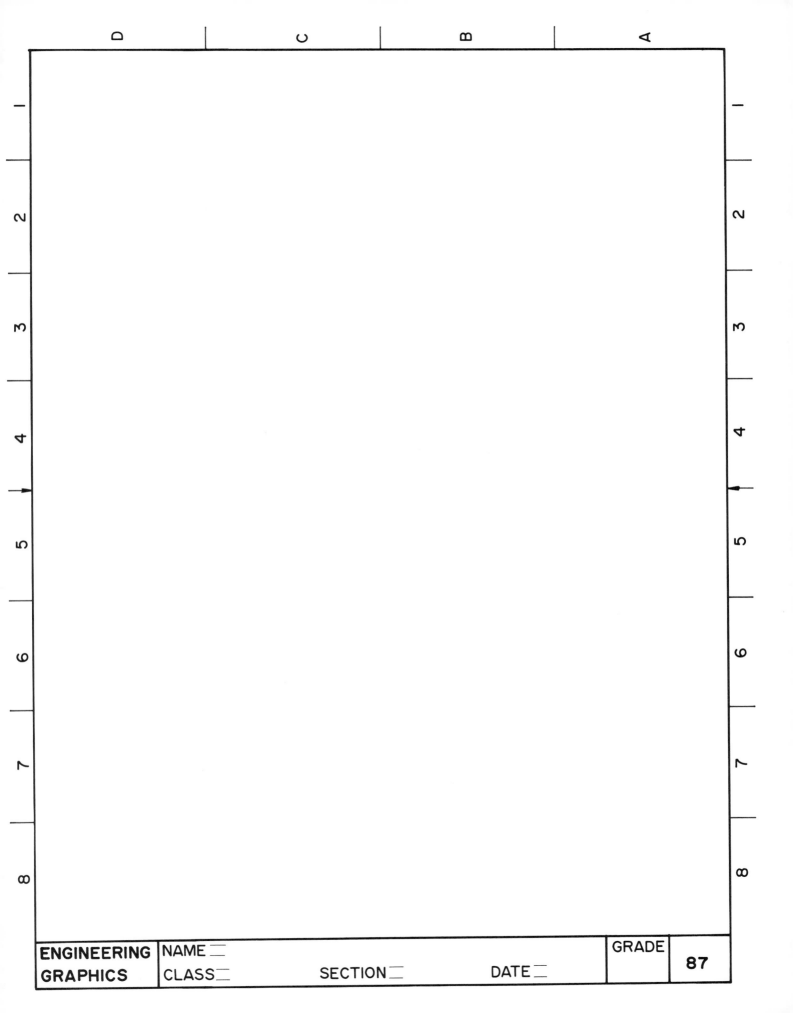

ENGINEERING GRAPHICS    NAME ▱    GRADE

CLASS ▱    SECTION ▱    DATE ▱    **87**

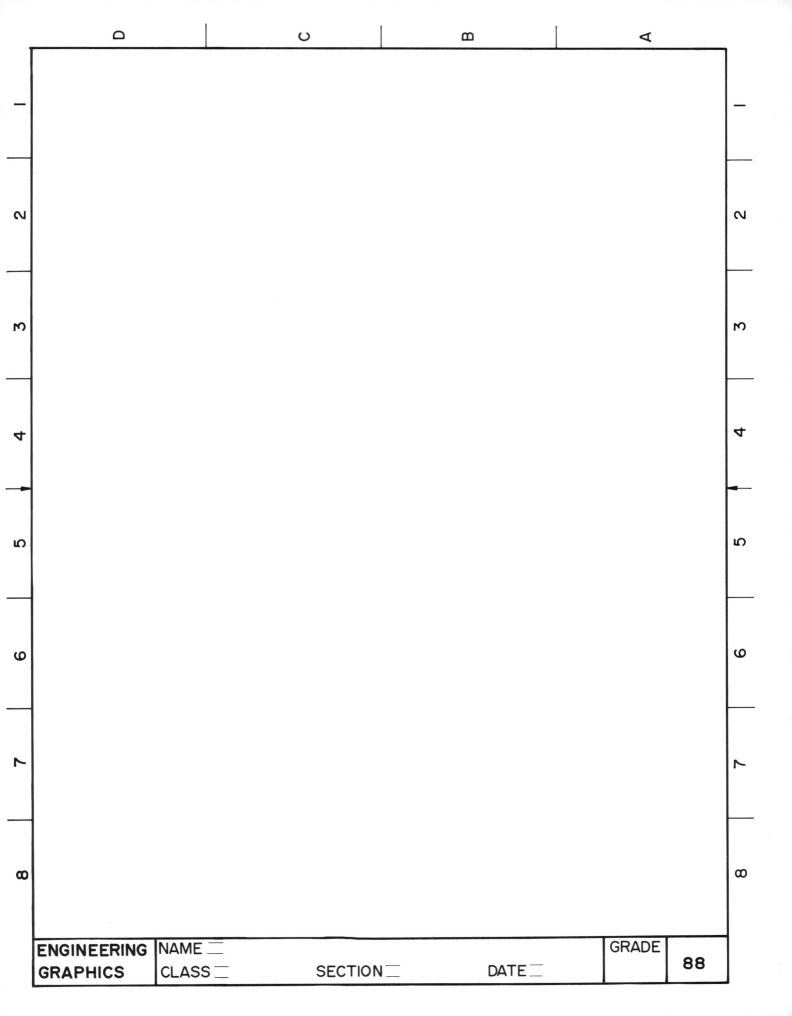

| D | C | B | A | | |
|---|---|---|---|---|---|
| **ENGINEERING** | NAME | | | GRADE | |
| **GRAPHICS** | CLASS SECTION DATE | | | | **88** |

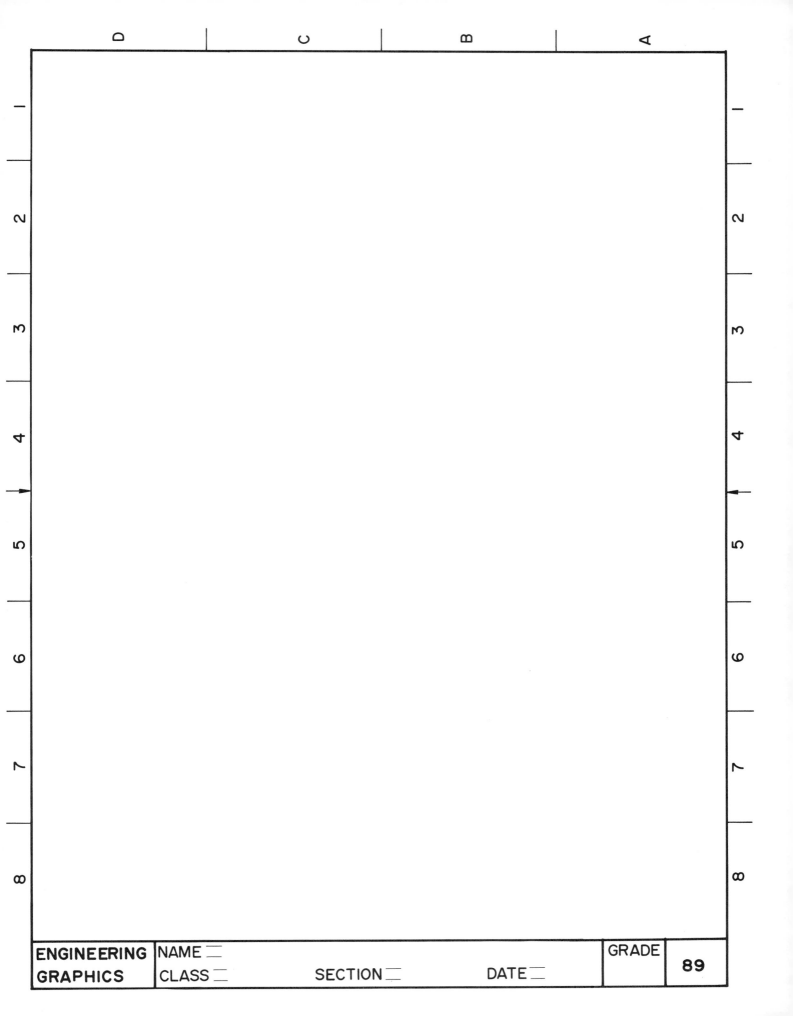

ENGINEERING
GRAPHICS

NAME
CLASS    SECTION    DATE

GRADE    89

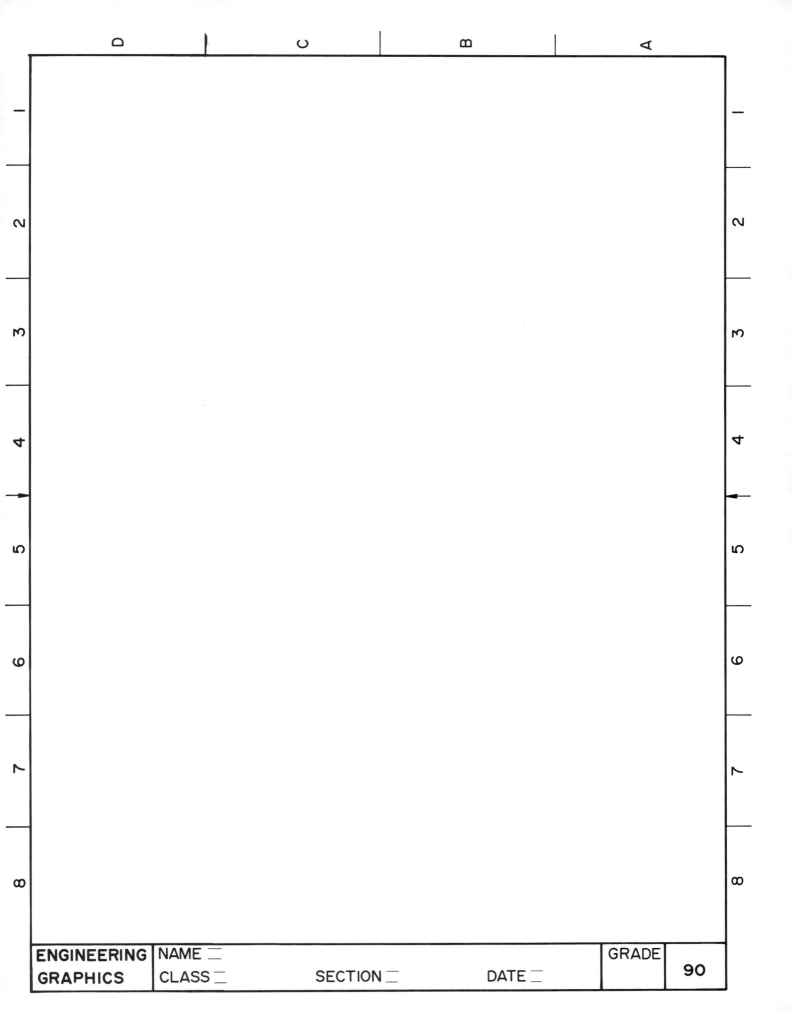

ENGINEERING GRAPHICS | NAME ☰ | CLASS ☰ | SECTION ☰ | DATE ☰ | GRADE | 90

MAKE A CHECK ASSEMBLY TO VERIFY THE DIMENSIONS ON
THIS PRODUCT. SEVERAL DIMENSIONS HAVE BEEN PURPOSELY
INCORRECTLY SPECIFIED. FIND THESE AND NOTE THE
CORRECTED SIZE ON THE PICTORIAL DRAWING.

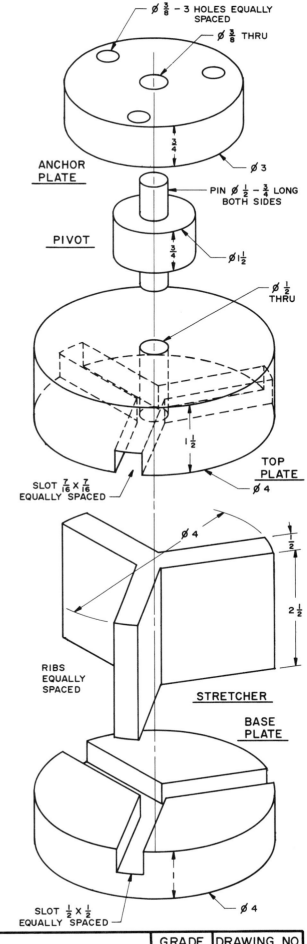

Ø $\frac{3}{8}$ – 3 HOLES EQUALLY
SPACED

Ø $\frac{3}{8}$ THRU

ANCHOR
PLATE

Ø 3

$\frac{3}{4}$

PIN Ø $\frac{1}{2}$ – $\frac{3}{4}$ LONG
BOTH SIDES

PIVOT

$\frac{3}{4}$

Ø $1\frac{1}{2}$

Ø $\frac{1}{2}$
THRU

$1\frac{1}{2}$

TOP
PLATE
Ø 4

SLOT $\frac{7}{16}$ X $\frac{7}{16}$
EQUALLY SPACED

Ø 4

$\frac{1}{2}$

$2\frac{1}{2}$

RIBS
EQUALLY
SPACED

STRETCHER

BASE
PLATE

SLOT $\frac{1}{2}$ X $\frac{1}{2}$
EQUALLY SPACED

Ø 4

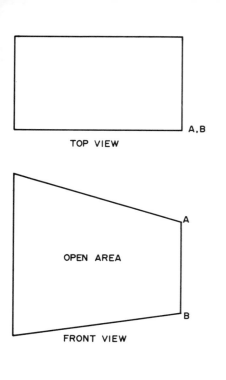

TOP VIEW

OPEN AREA

FRONT VIEW

A,B

A

B

1. LAY OUT THE INSIDE PATTERN OF THIS RECTANGULAR PIPE SECTION. ADD A $\frac{1}{4}$" WIDE TAB ALONG END A-B FOR ASSEMBLYING THE PIPE.

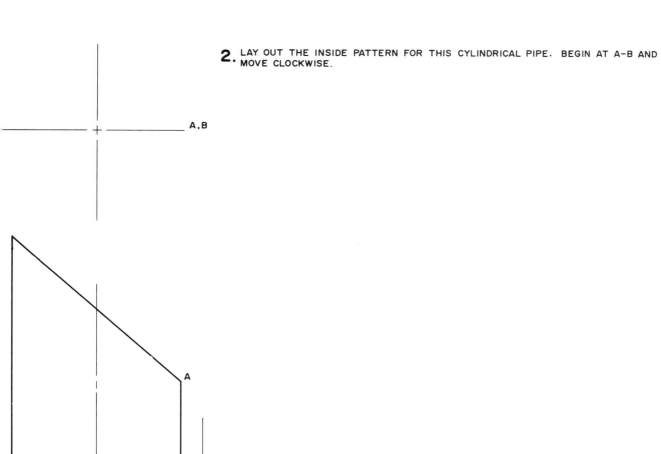

2. LAY OUT THE INSIDE PATTERN FOR THIS CYLINDRICAL PIPE. BEGIN AT A-B AND MOVE CLOCKWISE.

A,B

A

B

**DEVELOPMENTS**

| **ENGINEERING** | NAME ▭ | | | GRADE | DRAWING NO. |
| **GRAPHICS** | CLASS ▭ | SECTION ▭ | DATE ▭ | | **92** |

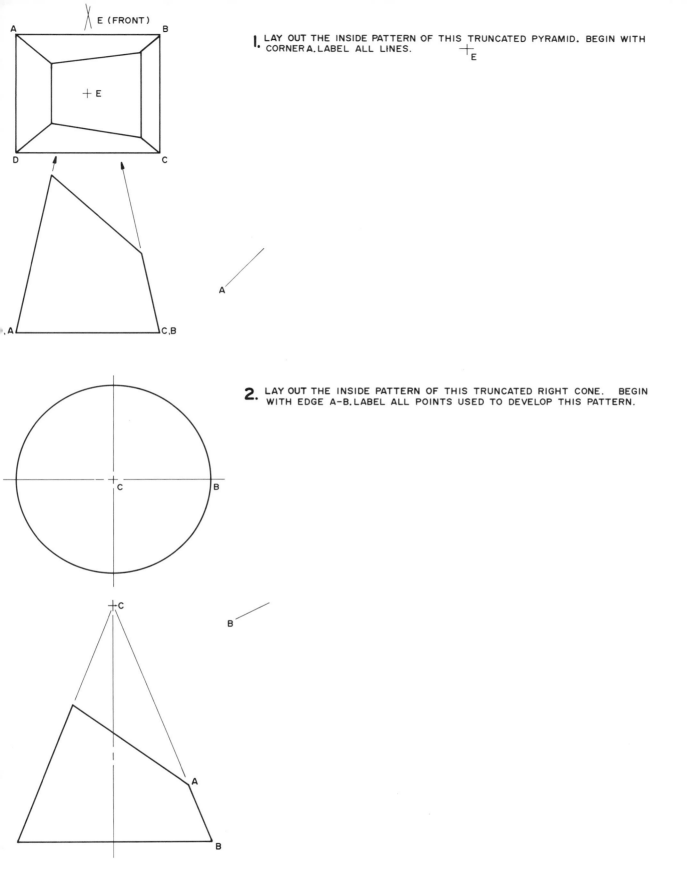

1. LAY OUT THE INSIDE PATTERN OF THIS TRUNCATED PYRAMID. BEGIN WITH CORNER A. LABEL ALL LINES.   +E

2. LAY OUT THE INSIDE PATTERN OF THIS TRUNCATED RIGHT CONE.   BEGIN WITH EDGE A-B. LABEL ALL POINTS USED TO DEVELOP THIS PATTERN.

## DEVELOPMENTS

| ENGINEERING | NAME ⎓ | | | GRADE | DRAWING NO. |
|---|---|---|---|---|---|
| GRAPHICS | CLASS ⎓ | SECTION ⎓ | DATE ⎓ | | 93 |

**I.** LAY OUT HALF OF THE PATTERN FOR THIS TRUNCATED OBLIQUE CONE. BEGIN THE LAYOUT USING ELEMENT A-B. LABEL ALL POINTS USED IN THE CONSTRUCTION AND ON THE FINISHED PATTERN.

B                                                      A
+————————————————————————————————————————+

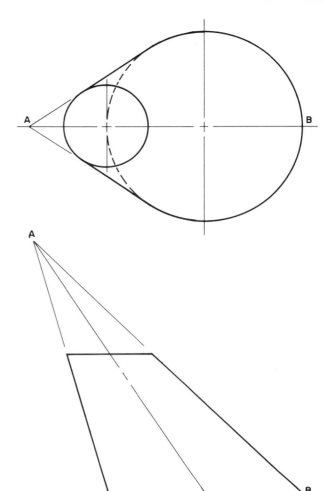

**2.** MAKE AN APPROXIMATE LAYOUT FOR HALF OF THE SPHERE BELOW USING THE CONE METHOD. LABEL ALL POINTS USED TO DEVELOP THE PATTERN. THE VERTEX WILL FALL ON TOP OF PROBLEM I.

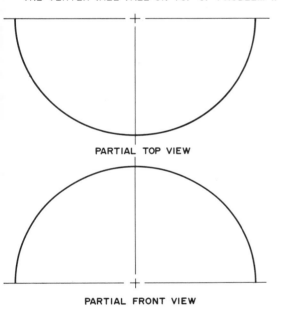

PARTIAL TOP VIEW

PARTIAL FRONT VIEW

## DEVELOPMENTS

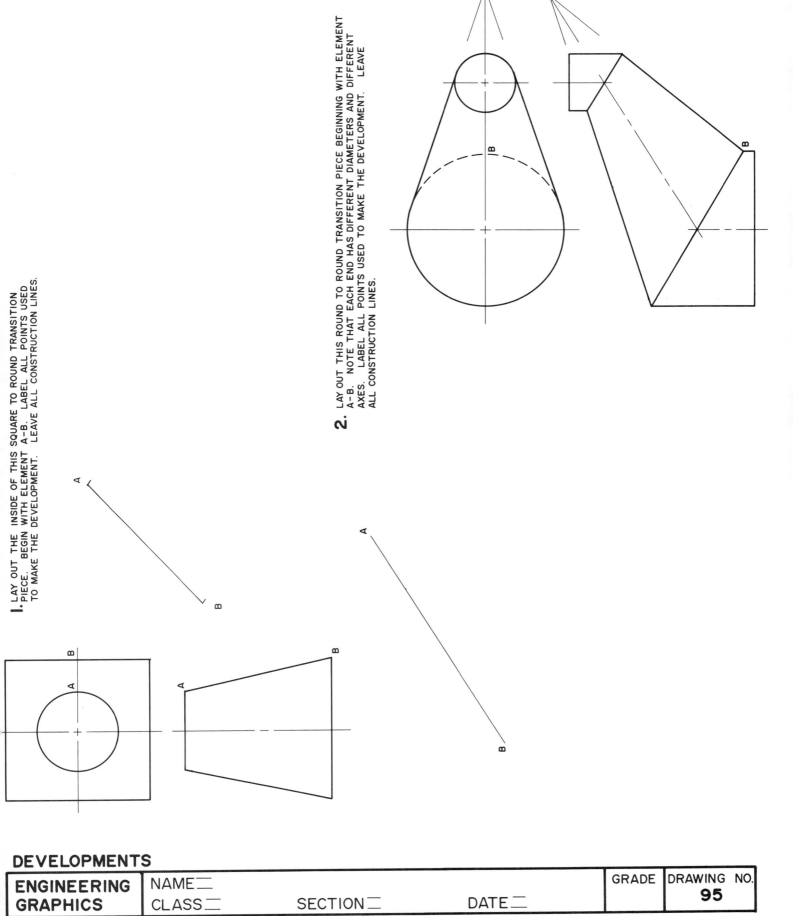

1. LAY OUT THE INSIDE OF THIS SQUARE TO ROUND TRANSITION PIECE. BEGIN WITH ELEMENT A-B. LABEL ALL POINTS USED TO MAKE THE DEVELOPMENT. LEAVE ALL CONSTRUCTION LINES.

2. LAY OUT THIS ROUND TO ROUND TRANSITION PIECE BEGINNING WITH ELEMENT A-B. NOTE THAT EACH END HAS DIFFERENT DIAMETERS AND DIFFERENT AXES. LABEL ALL POINTS USED TO MAKE THE DEVELOPMENT. LEAVE ALL CONSTRUCTION LINES.

# DEVELOPMENTS

| ENGINEERING GRAPHICS | NAME ▭ | | | GRADE | DRAWING NO. |
| --- | --- | --- | --- | --- | --- |
| | CLASS ▭ | SECTION ▭ | DATE ▭ | | 95 |

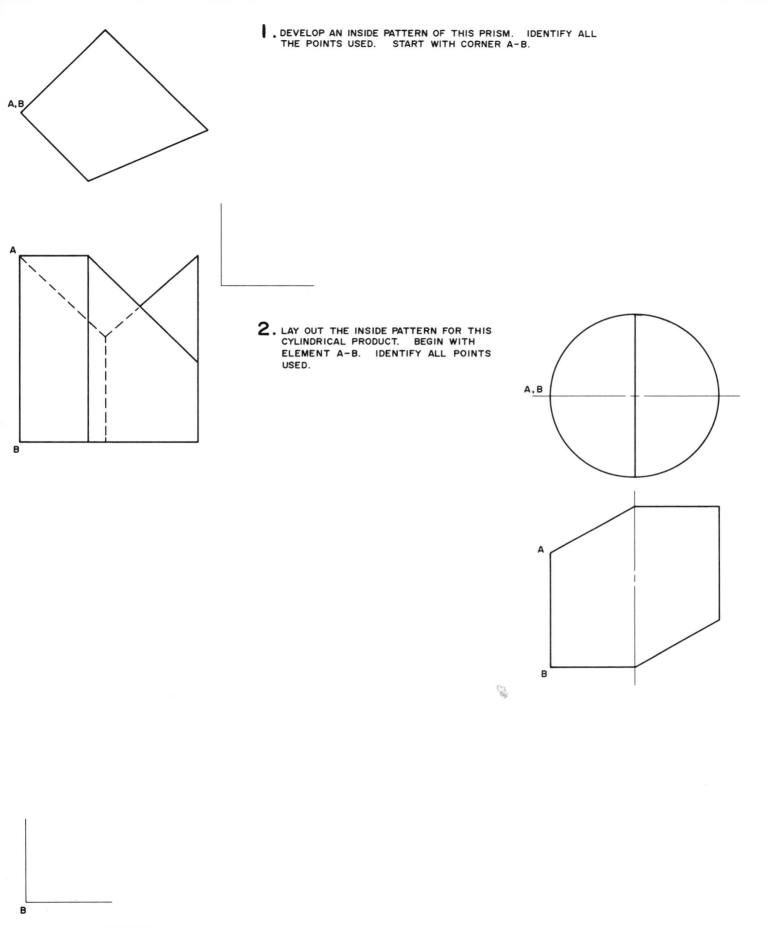

**1.** DEVELOP AN INSIDE PATTERN OF THIS PRISM. IDENTIFY ALL THE POINTS USED. START WITH CORNER A-B.

A,B

A

B

**2.** LAY OUT THE INSIDE PATTERN FOR THIS CYLINDRICAL PRODUCT. BEGIN WITH ELEMENT A-B. IDENTIFY ALL POINTS USED.

A,B

A

B

B

# DEVELOPMENTS

| ENGINEERING | NAME ⎓ | | | GRADE | DRAWING NO. |
|---|---|---|---|---|---|
| GRAPHICS | CLASS ⎓ | SECTION ⎓ | DATE ⎓ | | **96** |

**1.** FIND THE LINE OF INTERSECTION BETWEEN THE INCLINED PLANE AND THE PRISM.

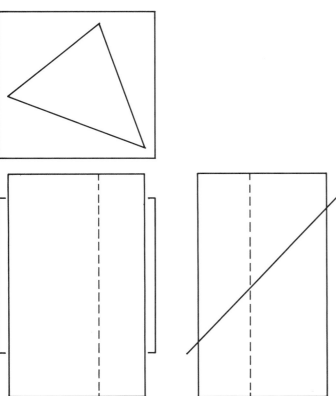

**2.** FIND THE LINE OF INTERSECTION BETWEEN THE OBLIQUE PLANE AND THE PRISM.

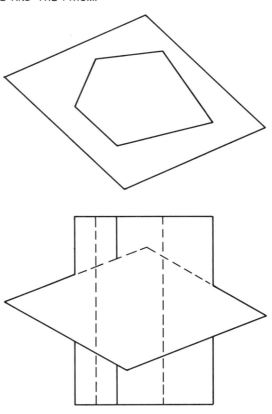

**3.** FIND THE LINE OF INTERSECTION BETWEEN THIS OBLIQUE PLANE AND THE RIGHT CYLINDER.

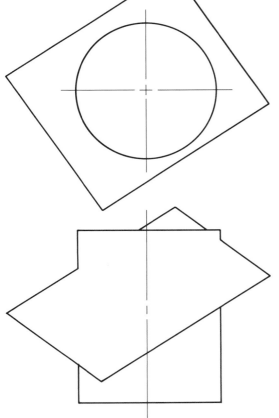

**4.** FIND THE LINE OF INTERSECTION BETWEEN THIS OBLIQUE PLANE AND THE RIGHT CYLINDER.

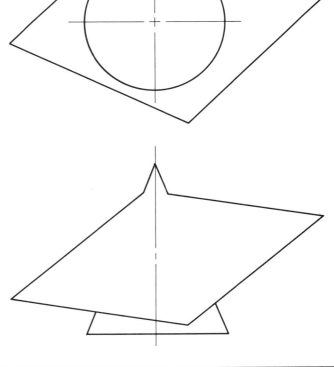

## INTERSECTIONS

| ENGINEERING | NAME | | | GRADE | DRAWING NO. |
| GRAPHICS | CLASS | SECTION | DATE | | **97** |

**1.** FIND THE LINE OF INTERSECTION BETWEEN THESE PRISMS.

**2.** FIND THE LINE OF INTERSECTION BETWEEN THESE PRISMS.

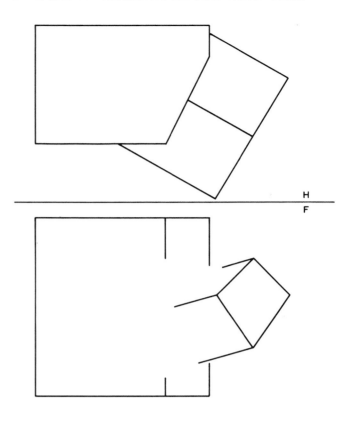

**3.** FIND THE LINE OF INTERSECTION BETWEEN THE OBLIQUE PRISM AND OBLIQUE PLANE.    DETERMINE VISIBILITY.

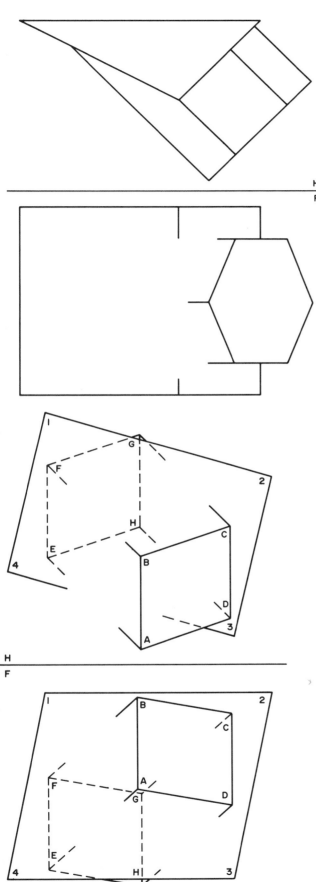

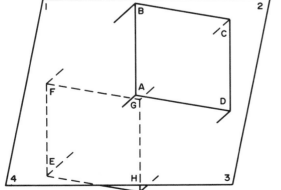

## INTERSECTIONS

| ENGINEERING | NAME ▭ | | | GRADE | DRAWING NO. |
|---|---|---|---|---|---|
| GRAPHICS | CLASS ▭ | SECTION ▭ | DATE ▭ | | **98** |

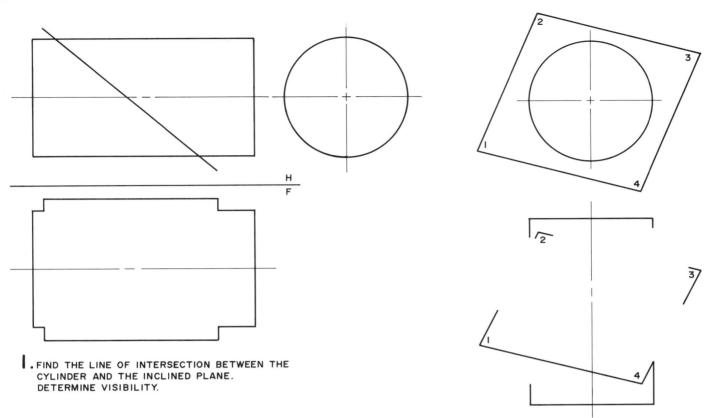

H
F

1. FIND THE LINE OF INTERSECTION BETWEEN THE
CYLINDER AND THE INCLINED PLANE.
DETERMINE VISIBILITY.

2. FIND THE LINE OF INTERSECTION BETWEEN THE
CYLINDER AND THE OBLIQUE PLANE.
DETERMINE VISIBILITY.

3. FIND THE LINE OF INTERSECTION BETWEEN THE
OBLIQUE CYLINDER AND OBLIQUE PLANE 1-2-3-4.
DETERMINE VISIBILITY.

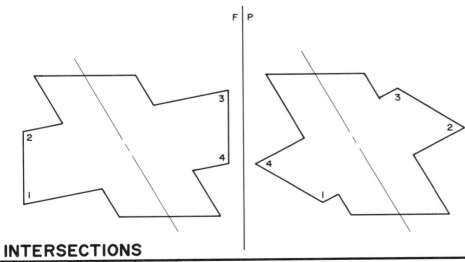

F | P

**INTERSECTIONS**

| **ENGINEERING** | NAME ⹀ | | | GRADE | DRAWING NO. |
|---|---|---|---|---|---|
| **GRAPHICS** | CLASS ⹀ | SECTION ⹀ | DATE ⹀ | | **99** |

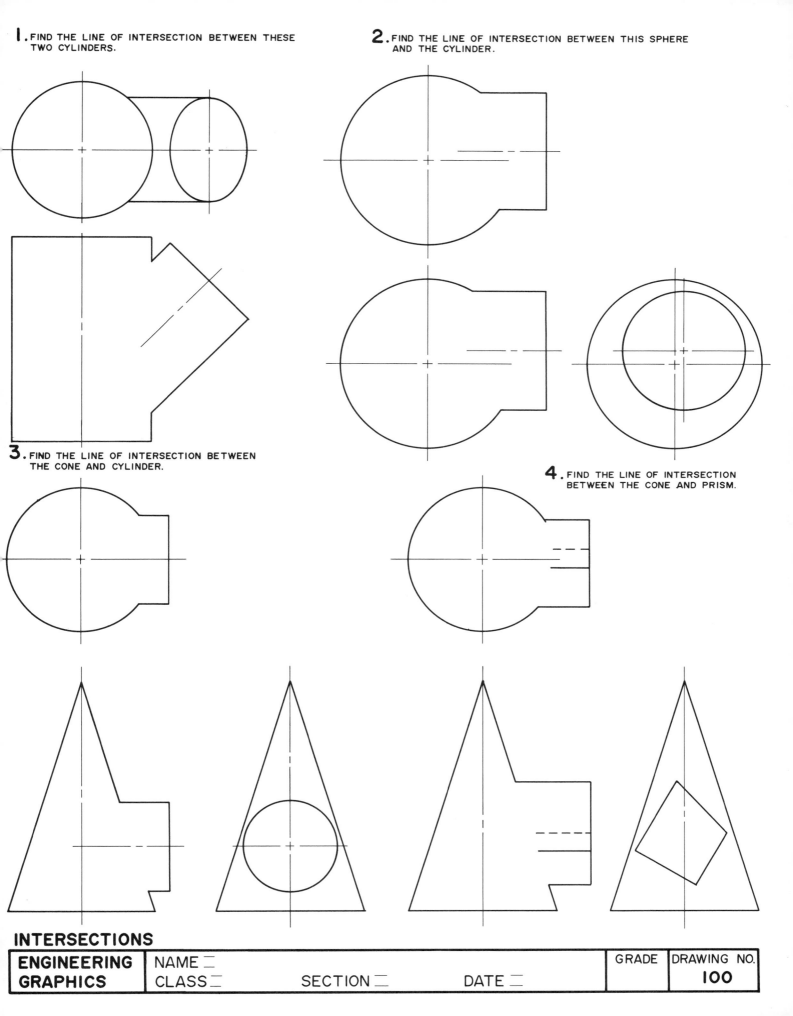

**1.** FIND THE LINE OF INTERSECTION BETWEEN THESE TWO CYLINDERS.

**2.** FIND THE LINE OF INTERSECTION BETWEEN THIS SPHERE AND THE CYLINDER.

**3.** FIND THE LINE OF INTERSECTION BETWEEN THE CONE AND CYLINDER.

**4.** FIND THE LINE OF INTERSECTION BETWEEN THE CONE AND PRISM.

**INTERSECTIONS**

| ENGINEERING GRAPHICS | NAME = | | | GRADE | DRAWING NO. |
|---|---|---|---|---|---|
| | CLASS = | SECTION = | DATE = | | 100 |

**1.** FIND THE ACTUAL DIRECTION AND SPEED OF THE SHIP BELOW.   SCALE: 1" = 2 KNOTS.

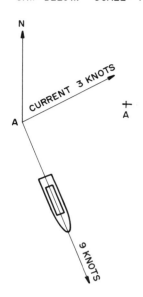

**2.** FIND THE FORCES IN THE SUPPORTS CARRYING THE WEIGHT BELOW.   SCALE: 1" = 100 POUNDS.

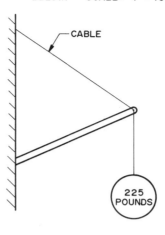

**3.** FIND THE RESULTANT OF THESE THREE CONCURRENT, COPLANAR FORCES USING THE PARALLELOGRAM AND THE POLYGON METHODS. MEASURE THE RESULTANTS AT THE SCALE: 1 mm = 5 N.

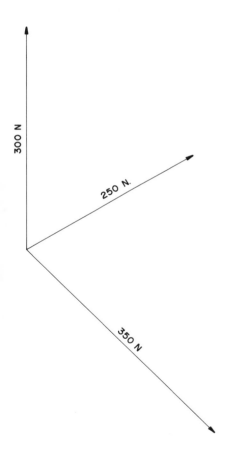

PARALLELOGRAM METHOD

POLYGON METHOD

## VECTOR ANALYSIS

| ENGINEERING GRAPHICS | NAME ⎓ CLASS ⎓ | SECTION ⎓ | DATE ⎓ | GRADE | DRAWING NO. 101 |
|---|---|---|---|---|---|

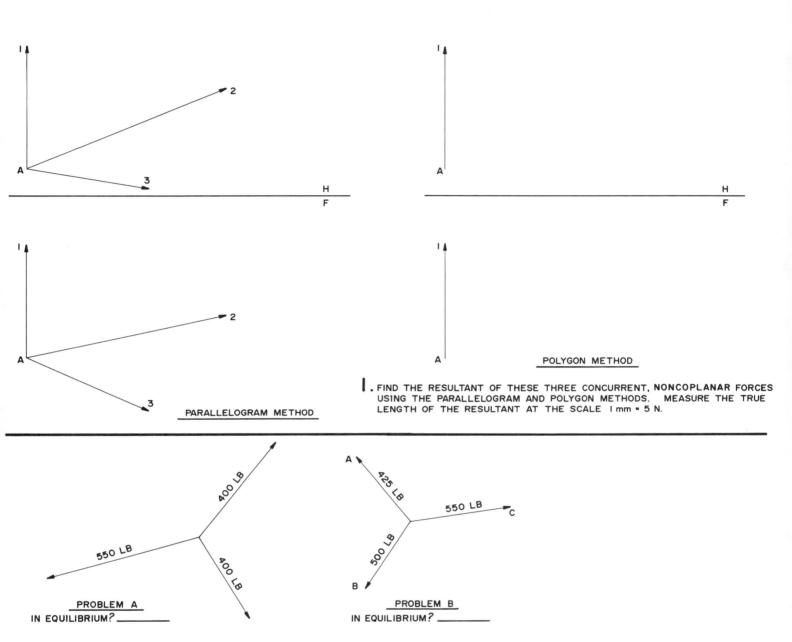

2

3

A

H
F

PARALLELOGRAM METHOD

POLYGON METHOD

I. FIND THE RESULTANT OF THESE THREE CONCURRENT, NONCOPLANAR FORCES
USING THE PARALLELOGRAM AND POLYGON METHODS.   MEASURE THE TRUE
LENGTH OF THE RESULTANT AT THE SCALE 1 mm = 5 N.

400 LB

550 LB

400 LB

PROBLEM A
IN EQUILIBRIUM? _____

A

425 LB

550 LB

C

500 LB

B

PROBLEM B
IN EQUILIBRIUM? _____

2. WHICH OF THESE CONCURRENT, COPLANAR FORCES ARE IN EQUILIBRIUM?
DRAW TO THE SCALE 1" = 200 LB.

**VECTOR ANALYSIS**

| ENGINEERING | NAME ⌐ | | | GRADE | DRAWING NO. |
|---|---|---|---|---|---|
| GRAPHICS | CLASS ⌐ | SECTION ⌐ | DATE ⌐ | | 102 |

1. FIND THE LIFTING CAPACITY OF THE CRANE WITH THE HOOK IN THE THREE POSITIONS SHOWN.
   SCALE: 1" = 750 LB.

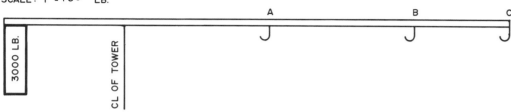

2. FIND THE MAGNITUDE, DIRECTION AND LINE OF ACTION OF A FORCE THAT WILL PLACE THESE NONCONCURRENT, COPLANAR FORCES
   IN EQUILIBRIUM.   SCALE: 1" = 100 LB.

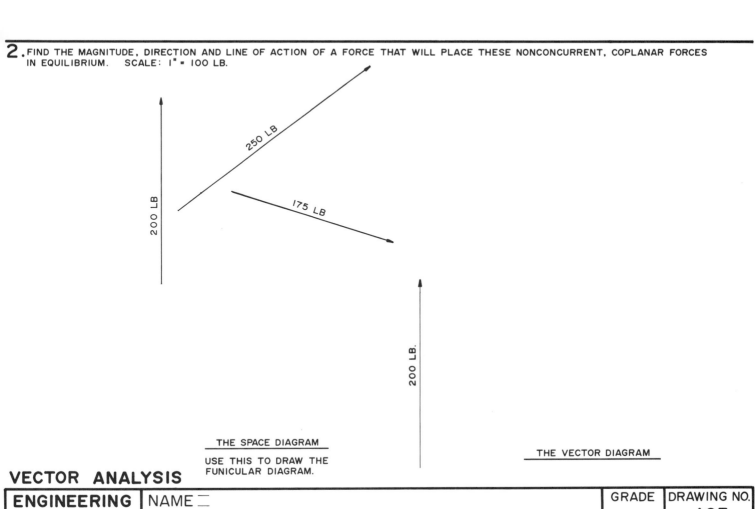

THE SPACE DIAGRAM

USE THIS TO DRAW THE
FUNICULAR DIAGRAM.

THE VECTOR DIAGRAM

**VECTOR ANALYSIS**

| ENGINEERING | NAME ▭ | | | GRADE | DRAWING NO. |
|---|---|---|---|---|---|
| GRAPHICS | CLASS ▭ | SECTION ▭ | DATE ▭ | | 103 |

**1.** FIND THE MAGNITUDE OF THE RESULTANT AT EACH END OF THE BEAM BELOW.
FIND THE POINT WHERE THE MAGNITUDE OF THE RESULTANT IS LOCATED.
SCALE: 1" = 240 LB.

360 LB.          240 LB.          480 LB.

A            B            C            D

R₁          SPACE DIAGRAM          R₂
USE THIS TO DRAW THE
FUNICULAR DIAGRAM.

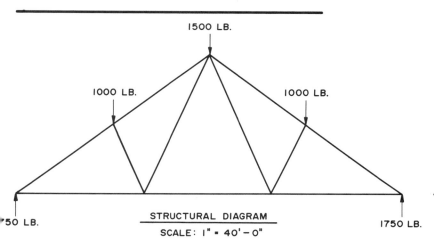

1500 LB.

1000 LB.                    1000 LB.

THE VECTOR DIAGRAM

STRUCTURAL DIAGRAM
50 LB.                                    1750 LB.
SCALE: 1" = 40'-0"

**2.** BEGINING WITH JOINT 1 DRAW A FREE BODY DIAGRAM
FOR EACH JOINT OF THE TRUSS AT THE LEFT.  THEN
CONSTRUCT A MAXWELL DIAGRAM AT THE SCALE
1" = 1000 LB.

+ ———————— —            ——— + ———

JOINT 1                    JOINT 3

+

JOINT 2

**TRUSS ANALYSIS**

| **ENGINEERING** | NAME ═ | | | GRADE | DRAWING NO. |
| **GRAPHICS** | CLASS ═ | SECTION ═ | DATE ═ | | 104 |

**1.** LOCATE THE DISPLACEMENT AND PATH OF POINT B AS POINT C ON THE SLIDER REACHES MINIMUM AND MAXIMUM DISTANCES FROM POINT A. IF SLIDER C MOVES 38 mm IN 1 SECOND WHAT IS THE AVERAGE VELOCITY? NOTE: ADD A PIECE OF PAPER ON THE LEFT.

**2.** ROTATE POINT B 360° COUNTERCLOCKWISE. LOCATE POINT C WHEN B HAS ROTATED 90°, 120°, 210° AND 300° FROM ITS POSITION. IF IT TAKES 3 SECONDS TO COMPLETE ONE 360° REVOLUTION WHAT IS THE ANGULAR VELOCITY OF POINT C?

AVERAGE VELOCITY _____

ANGULAR VELOCITY _____

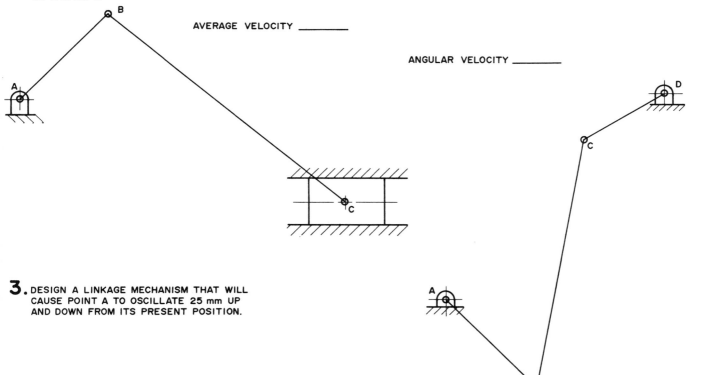

**3.** DESIGN A LINKAGE MECHANISM THAT WILL CAUSE POINT A TO OSCILLATE 25 mm UP AND DOWN FROM ITS PRESENT POSITION.

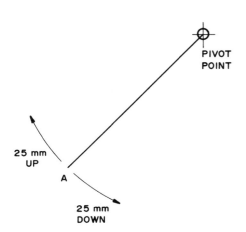

PIVOT POINT

25 mm UP

A

25 mm DOWN

# GRAPHICAL KINEMATICS

| **ENGINEERING** | NAME ⚊ | | | GRADE | DRAWING NO. |
|---|---|---|---|---|---|
| **GRAPHICS** | CLASS ⚊ | SECTION ⚊ | DATE ⚊ | | **105** |

DRAW THE DISPLACEMENT DIAGRAM AND THE CAM PROFILE FOR THE DATA GIVEN BELOW.
SCALE: FULL SCALE.

0°

360°

DISPLACEMENT DIAGRAM

| CAM DATA | |
|---|---|
| RISE – HARMONIC MOTION | 120° |
| DWELL – | 60° |
| FALL – HARMONIC MOTION | 180° |
| FOLLOWER DISPLACEMENT | 2.5" |
| BASE CIRCLE DIAMETER | 3.0" |
| OFFSET CIRCLE DIAMETER | 2.25" |
| KNIFE EDGE FOLLOWER | |
| HUB DIAMETER | 1.0" |
| CAM SHAFT DIAMETER | .62" |
| KEYWAY | .18" X .09" |

CENTER OF
CAM SHAFT

CAM PROFILE

**CAMS**

| ENGINEERING GRAPHICS | NAME ___<br>CLASS ___ SECTION ___ DATE ___ | GRADE | DRAWING NO. 106 |
|---|---|---|---|

THE FOLLOWING DATA ARE TO BE REPORTED IN CHART FORM.   DRAW THOSE ASSIGNED ON SHEETS 108, 109, 110.

**1.** SHOW THE ANNUAL INGOT PRODUCTION DATA BELOW A BAR OR PIE CHART.

| FURNACE TYPE | NET TONS |
|---|---|
| OPEN HEARTH | 88,950,000 |
| ELECTRIC | 11,650,000 |
| BASIC OXYGEN | 8,650,000 |
| BESSEMER | 975,000 |

**2.** MAKE A PIE CHART TO REPORT THE COMPOSITION OF EARTHLY MATTER.

| ELEMENT | PER CENT |
|---|---|
| SILICON | 26.0 |
| OXYGEN | 49.9 |
| ALUMINUM | 7.3 |
| IRON | 4.1 |
| CALCIUM | 3.2 |
| SODIUM | 2.3 |
| POTASSIUM | 2.3 |
| MAGNESIUM | 2.1 |
| ALL OTHER ELEMENTS | 2.8 |
| | 100.0 |

**3.** PREPARE A BAR CHART TO MAKE A COMPARISON OF MELTING TEMPERATURE OF THE MATERIALS BELOW.

| MATERIAL | TEMPERATURE °C |
|---|---|
| CARBON STEEL | 1540 |
| ALUMINUM ALLOYS | 648 |
| COPPER ALLOYS | 1090 |
| BRICK | 2760 |
| CONCRETE | 1650 |
| PLATE GLASS | 815 |
| POLYETHYLENE | 90 |
| EPOXIES | 260 |

**4.** REPORT THE DATA BELOW SHOWING THE AMOUNTS OF ALLOYING ELEMENTS USED IN VARIOUS STEEL ALLOYS. USE A RANGE BAR CHART.

| ELEMENT | PER CENT |
|---|---|
| MANGANESE | .50 to 2.0 |
| CHROMIUM | .50 to 12.0 |
| NICKEL | 1.0 to 27.0 |
| TUNGSTEN | 17.0 to 20.0 |
| MOLYBDENUM | .10 to 4.0 |
| COBALT | 17.0 to 36.0 |

**5.** REPORT THE DATA BELOW COMPARING THE NOISE REDUCTION VALUES OF EXTERIOR CONSTRUCTIONS.

| TYPE OF CONSTRUCTION | DIFFERENCE BETWEEN INSIDE AND OUTSIDE NOISE LEVELS IN DECIBELS. |
|---|---|
| $\frac{1}{4}$" PLATE GLASS WINDOW. | 11 to 13.5 |
| ALUMINUM FRAME WINDOW WITH $\frac{1}{4}$" GLASS AND AIR SPACE. | 35 to 41 |
| WOOD SHEATHED EXTERIOR WALL 2" X 4" STUDS, $\frac{1}{2}$" GYPSUM BOARD INSIDE. | 23.5 to 28 |
| 9" BRICK WALL | 41 to 46.5 |
| BUILT-UP ROOF WITH $\frac{1}{2}$" GYPSUM BOARD CEILING. | 36 to 41.5 |

**6.** PREPARE A CHART TO COMPARE THE COMPRESSIVE STRENGTH OF THE FOLLOWING CLASSES OF GRAY CAST IRON.

| ASTM CLASS | COMPRESSIVE STRENGTH, PSI. |
|---|---|
| 20 | 80,000 |
| 30 | 115,000 |
| 35 | 130,000 |
| 40 | 150,000 |
| 50 | 170,000 |
| 60 | 190,000 |

**8.** COMPARE THE BOILING POINTS OF THE FOLLOWING MATERIALS.

| MATERIAL | BOILING POINT, °F |
|---|---|
| ARGON GAS | -365 |
| OXYGEN | -360 |
| NITROGEN | -383 |
| DIAMONDS (CARBON) | +8500 |
| ALUMINUM | +4380 |
| IRON - CARBON STEEL | +5500 |

**9.** PREPARE A CHART TO SHOW THE RELATIONSHIP OF THE TENSILE STRENGTH OF THE FOLLOWING MATERIALS.

| MATERIAL | TENSILE STRENGTH, PSI |
|---|---|
| NATURAL RUBBER | 3,000 |
| POLYSTYRENE | 7,000 |
| FOSTERITE | 10,000 |
| WINDOW GLASS | 10,000 |
| ALUMINUM | 13,000 |
| IRON - CARBON STEEL | 200,000 |
| HOT ROLLED STEEL | 55,000 |

**7.** DRAW A PICTORIAL CHART TO REPORT THE DATA BELOW ON AUTOMOBILE PRODUCTION.   DESIGN A SYMBOL TO REPRESENT THE VARIABLE, THE NUMBER OF AUTOMOBILES.

| YEAR | NUMBER OF AUTOMOBILES |
|---|---|
| A | 3,750,500 |
| B | 4,150,250 |
| C | 4,350,000 |
| D | 3,900,500 |
| E | 3,700,200 |

**10.** PREPARE A FLOW CHART TO ILLUSTRATE HOW CONCENTRATED MAGNESIUM CHLORIDE IS EXTRACTED FROM OCEAN WATER.   FOLLOWING IS A DESCRIPTION OF THE PROCESS.   THE INITIAL INPUT INVOLVES PRODUCING LIME FROM OYSTER SHELLS.   THE SHELLS ARE TREATED IN A KILN AT 2400 °F.   SEAWATER AND THE LIME ARE MIXED IN A TANK.   MAGNESIUM HYDRATE SETTLES TO THE BOTTOM OF THE TANK. IT IS THEN FILTERED TO PRODUCE A MORE CONCENTRATED HYDRATE.   THIS IS CONVERTED TO MAGNESIUM CHLORIDE BY ADDING HYDROCHLORIC ACID.   WATER IS THEN REMOVED BY EVAPORATION.   SPECIAL DRYING AND FILTERING FOLLOW PRODUCING MAGNESIUM CHLORIDE HAVING A 68% CONCENTRATION.

**CHARTS AND GRAPHS**

| ENGINEERING GRAPHICS | NAME CLASS | SECTION          DATE | | GRADE | DRAWING NO. 107 |
|---|---|---|---|---|---|

PLOT THE ASSIGNED PROBLEM ON THE THREE – CYCLE LOGARITHMIC GRID BELOW.

1. PUMP OUTPUT IN RELATION TO REVOLUTIONS PER MINUTE.

| R.P.M. | GALLONS / MINUTE |
|--------|------------------|
| 50 | 5 |
| 100 | 10 |
| 150 | 17 |
| 200 | 25 |
| 250 | 32 |
| 300 | 39 |
| 350 | 47 |
| 400 | 52 |
| 450 | 55 |
| 500 | 57 |

2. TEMPERATURE VERSUS PRESSURE OF FREON GAS.

| TEMPERATURE °C | PRESSURE OF FREON REFRIGERANT – PSI |
|----------------|-------------------------------------|
| 1.0 | 30.1 |
| 10.0 | 46.7 |
| 15.0 | 57.7 |
| 20.0 | 67.6 |
| 25.5 | 81.3 |
| 30.0 | 93.3 |
| 35.5 | 110.0 |
| 40.0 | 124.6 |
| 45.5 | 144.7 |
| 50.0 | 162.2 |

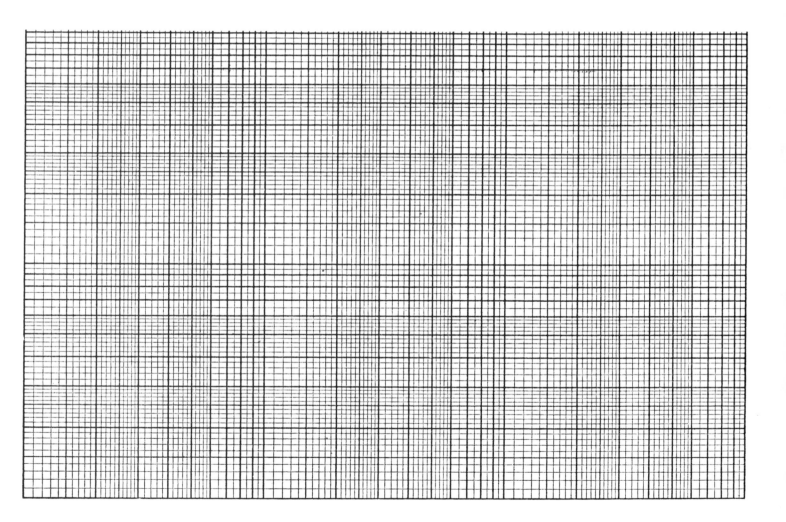

**LOGARITHMIC CHART**

| ENGINEERING GRAPHICS | NAME ‾  CLASS ‾        SECTION ‾        DATE ‾ | GRADE | DRAWING NO. 111 |

PLOT THE CURVE OF VISUAL SPEED VERSUS LUMINANCE SHOWN BY THE DATA BELOW.   THE CURVE SHOWS THAT BY INCREASING THE
LUMINANCE (BRIGHTNESS) THE SEEING TIME IS REDUCED. PLACE THE VISUAL SPEED DATA ON THE Y-AXIS AND LUMINANCE ON THE X-AXIS.
LUMINANCE IS IN FOOTLAMBERTS.

| SPEED OF VISION | LUMINANCE (FOOTLAMBERTS) | SPEED OF VISION | LUMINANCE (FOOTLAMBERTS) | SPEED OF VISION | LUMINANCE (FOOTLAMBERTS) |
|---|---|---|---|---|---|
| 0.120 | 1 | 0.320 | 20 | 0.372 | 200 |
| 0.170 | 2 | 0.340 | 30 | 0.375 | 300 |
| 0.200 | 3 | 0.345 | 40 | 0.378 | 400 |
| 0.220 | 4 | 0.350 | 50 | 0.380 | 500 |
| 0.235 | 5 | 0.355 | 60 | 0.383 | 600 |
| 0.245 | 6 | 0.360 | 70 | 0.385 | 700 |
| 0.255 | 7 | 0.365 | 80 | 0.387 | 800 |
| 0.260 | 8 | 0.367 | 90 | 0.390 | 900 |
| 0.275 | 9 | 0.370 | 100 | 0.392 | 1000 |
| 0.280 | 10 | | | | |

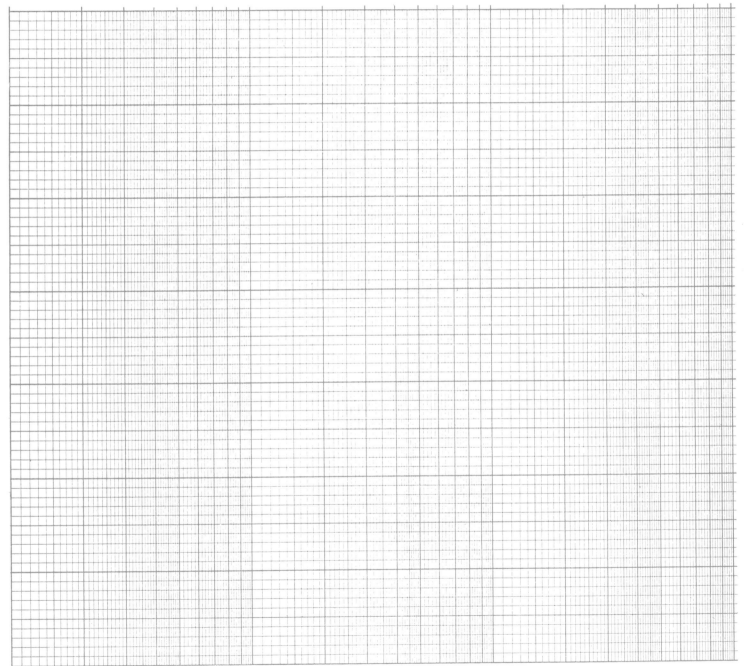

## SEMILOGARITHMIC CHART

| ENGINEERING GRAPHICS | NAME ⁼ CLASS ⁼ | SECTION ⁼ | DATE ⁼ | GRADE | DRAWING NO. 112 |
|---|---|---|---|---|---|

**I.** ON THE CHART AT THE RIGHT COMPARE THE RELATIVE HUMIDITY VALUES RECOMMENDED FOR RESIDENTIAL STRUCTURES FOR VARIOUS OUTDOOR TEMPERATURE LEVELS.

| SINGLE GLAZED WINDOWS | | DOUBLE GLAZED WINDOWS | | TRIPLE GLAZED WINDOWS | |
|---|---|---|---|---|---|
| RH – %[1] | °F[2] | RH – %[1] | °F[2] | RH – %[1] | °F[2] |
| 5 | –30 | 31 | –30 | 42 | –30 |
| 8 | –20 | 35 | –20 | 44 | –20 |
| 11 | –10 | 40 | –10 | 48 | –10 |
| 15 | 0 | 45 | 0 | 55 | 0 |
| 20.5 | 10 | 51 | 10 | 60 | 10 |
| 28 | 20 | 56 | 20 | 65 | 20 |
| 37 | 30 | 64 | 30 | 71 | 30 |
| 47 | 40 | 70 | 40 | 76 | 40 |

1. RELATIVE HUMIDITY IN PERCENT.

2. OUTDOOR TEMPERTURE IN °F.

**2.** ON THE CHART BELOW COMPARE THE RELATIVE ACOUSTICAL PERFORMANCE OF THE FOLLOWING THREE DOORS.

| SOLID CORE GASKETED DOOR | | HOLLOW CORE GASKETED DOOR | | UNGASKETED DOOR | |
|---|---|---|---|---|---|
| STL – Db[1] | CPS[2] | STL – Db[1] | CPS[2] | STL – Db[1] | CPS[2] |
| 28 | 125 | 21 | 125 | 12 | 125 |
| 31 | 250 | 26 | 250 | 17 | 250 |
| 36 | 500 | 29 | 500 | 19 | 500 |
| 38 | 1500 | 31 | 1500 | 18 | 1500 |
| 40 | 2000 | 34 | 2000 | 17 | 2000 |

1. SOUND TRANSMISSION CLASS IN DECIBELS.

2. FREQUENCY CYCLES PER SECOND.

# LINEAR CHARTS

| ENGINEERING GRAPHICS | NAME ⹀ CLASS ⹀ SECTION ⹀ DATE ⹀ | GRADE | DRAWING NO. 113 |
|---|---|---|---|

PLOT THE DATA BELOW ON THE THREE GRIDS SHOWN AND WRITE THE EQUATION FOR THE STRAIGHT LINE.

THE EQUATION DEVELOPED IS _____ .

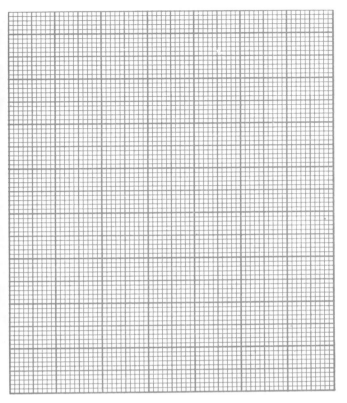

Y = MX + B

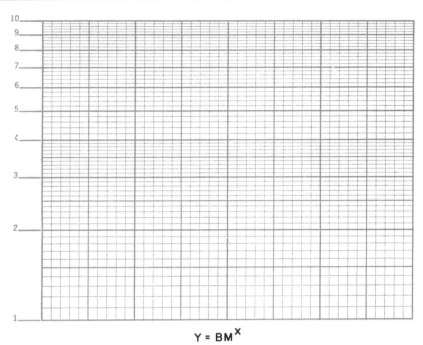

Y = BM$^X$

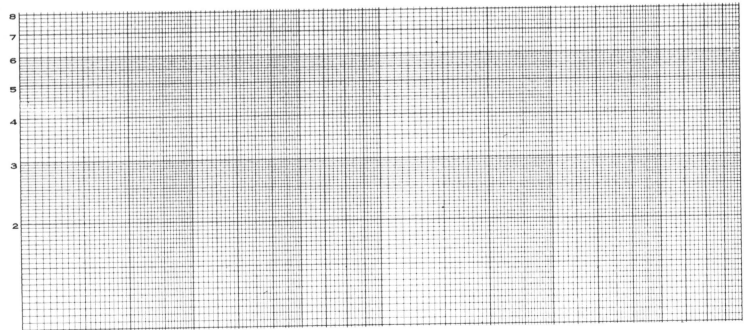

Y = BX$^m$

CONSTRUCT A LINEAR GRAPH TO DETERMINE THE EQUATION FOR THE COST OF
OPERATING A GASOLINE POWERED WATER PUMP.   THE DATA AT THE RIGHT
RELATES THE NUMBER OF GALLONS OF GASOLINE USED PER HOUR TO RUN THE
PUMP TO THE FLOW OF WATER IN GALLONS PER MINUTE.

| GASOLINE | WATER |
|----------|-------|
| .45 | 2 |
| 1.8 | 3 |
| 2.1 | 4 |
| 2.35 | 5 |
| 2.6 | 6 |
| 2.8 | 7 |

| GASOLINE | WATER |
|----------|-------|
| 3.0 | 8 |
| 3.2 | 9 |
| 3.4 | 10 |
| 4.9 | 20 |
| 6.0 | 30 |
| 7.0 | 40 |

## LINEAR EQUATIONS

| ENGINEERING | NAME ▭ | | | GRADE | DRAWING NO. |
|-------------|--------|---|---|-------|-------------|
| GRAPHICS | CLASS ▭ | SECTION ▭ | DATE ▭ | | 114 |

FIND THE TOTAL AREA OF THE SHEET
METAL PART BELOW USING GRAPHICAL
INTEGRATION.

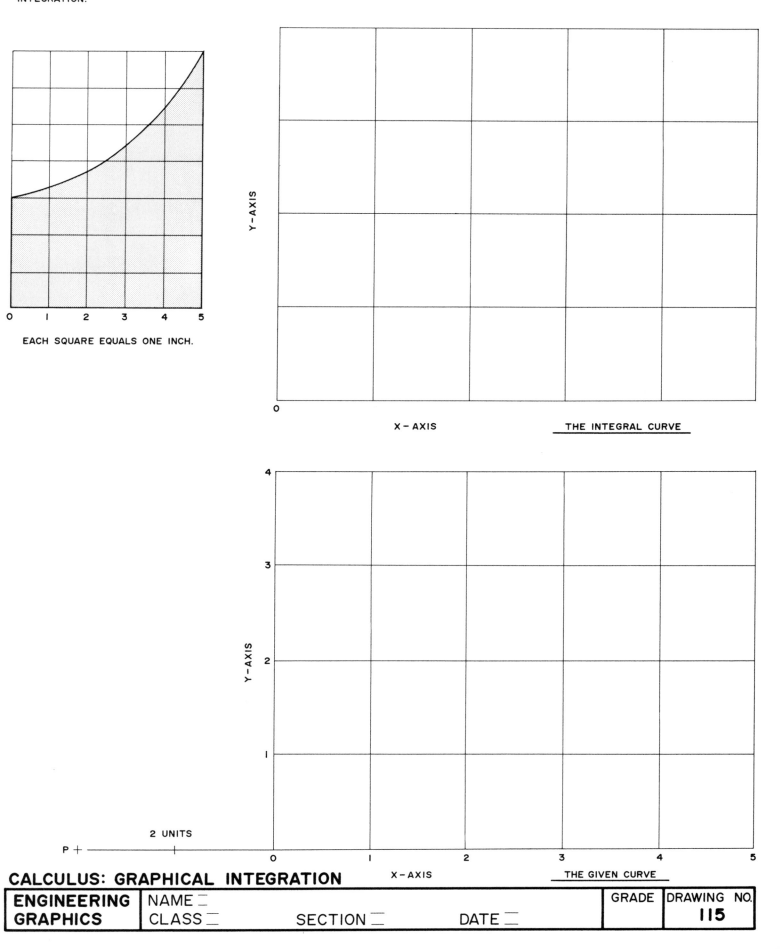

EACH SQUARE EQUALS ONE INCH.

Y-AXIS

O

X-AXIS          THE INTEGRAL CURVE

4

3

Y-AXIS

2

1

2 UNITS

P +

0          1          2          3          4          5

X-AXIS          THE GIVEN CURVE

**CALCULUS: GRAPHICAL INTEGRATION**

| **ENGINEERING** | NAME | | | GRADE | DRAWING NO. |
|---|---|---|---|---|---|
| **GRAPHICS** | CLASS | SECTION | DATE | | **115** |

PLOT THE EQUATION  X = $\frac{y^3}{2}$ AS A RECTANGULAR GRAPH.   THEN FIND THE
DERIVATIVE CURVE.   USE Y – VALUES 0 to 5.

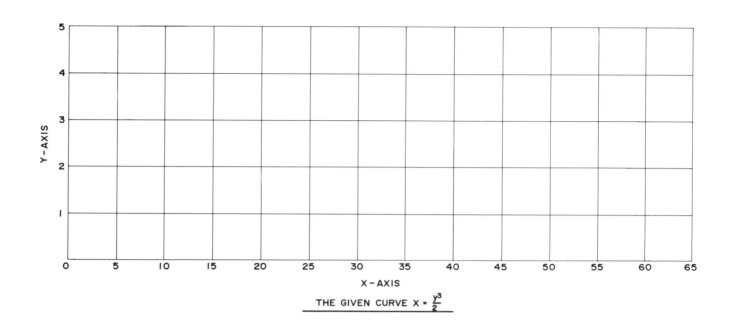

THE GIVEN CURVE X = $\frac{y^3}{2}$

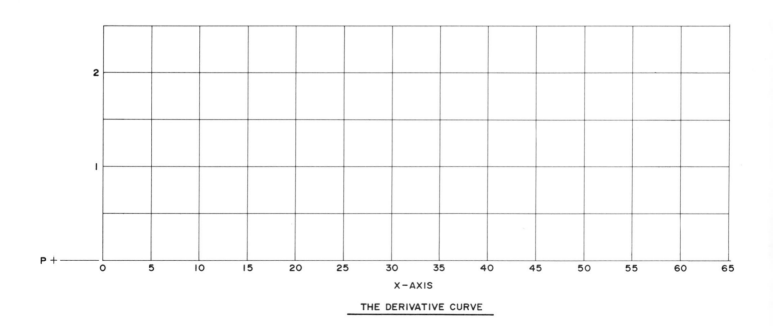

X–AXIS

THE DERIVATIVE CURVE

## CALCULUS: GRAPHICAL DIFFERENTIATION

| ENGINEERING | NAME ⁼ | | | GRADE | DRAWING NO. |
|---|---|---|---|---|---|
| GRAPHICS | CLASS ⁼ | SECTION ⁼ | DATE ⁼ | | 116 |

**1.** WRITE A PROGRAM TO DRAW A CROSS USING THE SPECIFIED COORDINATES FOR EACH END OF EACH LINE. THESE COORDINATES ARE : LINE NO. 1-30,80 AND 140,80; LINE NO. 2-70,20 AND 70,135. PLOT THESE ON THE GRID BELOW.

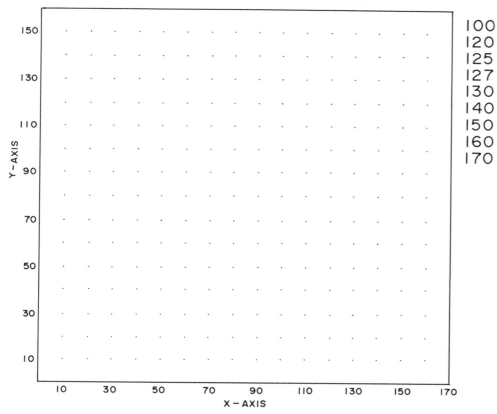

100
120
125
127
130
140
150
160
170

**2.** WRITE A PROGRAM TO DRAW A RECTANGLE USING THE SPECIFIED COORDINATES FOR EACH CORNER. USE MOVE AND DRAW STATEMENTS. THE COORDINATES ARE 30, 30; 30,110; 150,110; 150,30. PLOT THESE ON THE GRID BELOW.

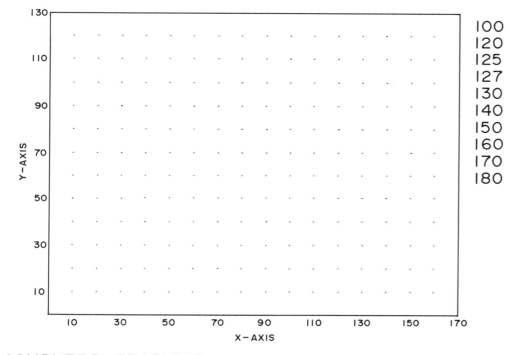

100
120
125
127
130
140
150
160
170
180

# COMPUTER GRAPHICS

| ENGINEERING | NAME = | | | | GRADE | DRAWING NO. |
|---|---|---|---|---|---|---|
| GRAPHICS | CLASS = | SECTION = | DATE = | | | 117 |